AFRO-OPTIMISM

Perspectives on Africa's Advances

Edited by
Ebere Onwudiwe and Minabere Ibelema

PRAEGER

Westport, Connecticut
London

Library of Congress Cataloging-in-Publication Data

Afro-optimism : perspectives on Africa's advances / edited by Ebere Onwudiwe and Minabere Ibelema.

p. cm.

Includes bibliographical references and index.

ISBN 0–275–97586–X (alk. paper)

1. Africa—Civilization—20th century. 2. Postcolonialism—Africa. 3. Africa—Economic conditions—1960– 4. Political culture—Africa. 5. Infrastructure (Economics)—Africa. 6. Public health—Africa. 7. Agriculture—Africa. I. Onwudiwe, Ebere, 1952– II. Ibelema, Minabere, 1954–

DT14.A418 2003
306′.096—dc21 2002025308

British Library Cataloguing in Publication Data is available.

Library of Congress Catalog Card Number: 2002025308
ISBN: 0–275–97586–X

First published in 2003

Praeger Publishers, 88 Post Road West, Westport, CT 06881
An imprint of Greenwood Publishing Group, Inc.
www.praeger.com

Printed in the United States of America

The paper used in this book complies with the Permanent Paper Standard issued by the National Information Standards Organization (Z39.48–1984).

10 9 8 7 6 5 4 3 2 1

To our children,
respectively, Chinwe, Memme, Awo,
and
Boma, and Ibim,
for whom Afro-optimism is most significant

Contents

Preface

This book crystallizes Black Africa's successes since political independence. There is a need to discuss Africa's achievements, to recognize her positives, and to make them a part of the general academic and popular discourses.

The necessity to do this, of course, says more about the scholarship on and media portrayal of Africa than it says about Africa. But this need is neither here nor there. The deprecation of Africa is widespread and enduring. It dates to precolonial scholarship and reporting on Africa and abounds in contemporary works. It permeates news coverage as well as television and cinematic portrayal. This volume is intended to be a counterpoint to this near universal portrayal and perception—what the then U.S. assistant secretary of state for African affairs, Chester Crocker, once decried as Afro-pessimism. Here we make the case for Afro-optimism.

There are some works that have illuminated Africa's pre-colonial glories, but the view that sub-Saharan Africa has regressed since independence from colonial powers is one that has hardly been challenged. We challenge it here. The chapters in this volume examine a wide range of elements of societal advancement and conclude that sub-Saharan Africa has made progress in all facets, including (gasp!) the economic. Evidently, our view of development is that it encompasses much more than the economic.

We are not under illusions regarding Africa's failures and problems, nor have we tried to whitewash the realities. Rather, we provide what is missing in the discourse in hopes of moving it toward a balanced perspective. We believe that such a balance is important to Africa and to the world at large.

We do not dare to think that a work of this nature will undo centuries of distortion. But we hope that several of the chapters force readers to reflect on Africa's other stories—the success stories.

Part I

PERSPECTIVE

1

Introduction: A Context for Post-Colonial African Discourse

Ebere Onwudiwe

Much of the current writing on post-colonial Africa has drawn a picture of a continent whose underdevelopment has been exacerbated by self-rule since the early 1960s. Independence, according to this dominant view, gave rise to monstrous corruption and mismanagement, which aggravated poverty and reduced the quality of life below pre-colonial levels. This long-standing view of Africans as inherently inferior and incapable of self-rule, made fashionable in imperial and colonial times, has received significant boosts in recent times not just because of the civil wars that have visited a few of over 54 African countries but more significantly from critics such as Ayittey (1992), Johnson (1992, 1993), and Kaplan (1994).[1] One common theme in both the academy and the media is the denunciation of African post-colonial leaders. This book shows that Africa has made more progress than is suggested by these and other critics and by Africa's dismal image.

A work of this nature is important for intellectual, economic, and other reasons. There is an intellectual necessity to describe Africa's reality in ways that counter perpetual inaccuracies. In this book, we are concerned with knowledge for its own sake. There are also economic ramifications for Africa's skewed image. To the extent that Africa's problems embody its image and its successes are overlooked, the continent will continue to be an investment desert to international businesses, much as American inner cities have been to American businesses.

In a world that has become intricately interdependent, few regions can thrive outside the economic lifeline of the global economy, and those with the worst images are sure to come up short in the bargain. There is, therefore, a moral imperative to rectify distortions that have the effect of further marginalizing the marginalized. We do not intend to ignore Africa's problems or to deny internal responsibility. Our goal is to tell Africa's other story, to make the case that Africa's successes to date are bases for a hopeful future if detrimental internal tendencies and external factors are reversed. We believe that the academy is one necessary place to begin the reversal.

The award-winning book *The Invention of Africa* (Mudimbe, 1988) has authoritatively demonstrated that the scholarly perspective through which Africans and Africanists consider Africa's identity lies in the academic discourses originating from both western missionary and anthropological myths about Africa. On occasion, the many dehumanizing myths about black people which form the core historical element in the traditional western definition of Africa (Thompson, 1985; Mudimbe, 1988) have been countered by politicians such as the Reverend Jesse Jackson with elegant slogans such as, "I am Somebody;" by musicians such as James Brown with, "I'm Black and Proud;" and by black scholars who proffer alternative constructs such as African personality, negritude, and pan-Africanism (Irele, 1990, Part 2). These counteractions are provoked by ceaseless popular and academic attacks on the African. The attacks are not limited to academic writers of the past, as Curtin (1995) would suggest. Current images of Africa are also mostly negative (Hawk, 1992). To get many black people, especially the younger ones, to believe in the above reaffirming slogans is to get them to go beyond the perceptions created by an unduly and pervasively negative portrayal.

European slave traders promoted and exploited the lore of a primitive and foreboding Africa as a defense of the primitive and inhuman trade. According to Hammond and Jablow (1970, p. 233), "African behavior, institutions, and character were not merely disparaged but presented as the negation of all human decencies." This "invention of Africa" became the basis of cyclical academic treatises of the type regurgitated in Herrnstein and Murray's (1994) book, *The Bell Curve.* Underlying such analysis is the belief that African abilities are race determined. Such claims often have political and economic impetus. Thompson (1985, p.14), for example, comments that throughout the period of European empires in Africa, "Scientific racism was a perfect fit with the global political economy period of white hegemony."

By codifying Western beliefs about Black Africa, academics generated degradation and other devaluing myths about the continent. The media have, in turn, propagated an image of Africa that is perennially negative. Can Africa live with this distortion of its image? Does the negative image of Africa in powerful countries such as the United States, for example, have development implications? Is it possible that there may be economic consequences that are positively related to the negative image of Black Africa?

We believe that such consequences exist and that the ability of a poor country to negotiate with a rich and powerful country has much to do with the relative power between the two countries (Onwudiwe, 1996). As argued in the work just cited, power is the global currency for achieving national interest in international relations, and the main ingredient in the power of the weak state is the image it has in the minds of decision makers in the powerful states, that is, states where power has many components.

Among the various definitions of power is one provided by Morganthau (1967, p. 26) which puts an emphasis on "man's control over the minds and actions of other men." Powerful countries employ economic, military, or diplomatic muscles as instruments of power. The poorest countries must depend on subtler and more intangible instruments of appeal. Therein lies the importance of a good or positive external image for the weak state.

For example, what does it mean to talk of the power of the United States in its relation with the state of Ghana? The power disparity is so great that the employment of tangible definitions of power as measures of the relation between the two countries can be misleading. Using such traditional measures (population, gross national product, military/industrial capacities, geographical location, etc.) alone, one can sensibly talk about the power of the United Sates in its relation to Ghana. But since Ghana does not have enough tangible power to influence the behavior of the United States through the threat of coercion, sanctions, or force, it makes little sense to talk about the power of Ghana vis-à-vis the United States. The same is true for all black African countries. The difference between the gross quantity of power of any black African country and that of the United States is simply too large to permit the meaningful use of this measure. For this reason, it may be the intangible component of power—including the appeal, image, or perception of Ghana in the United States—which is most available to Ghana to use in its dealings with the United States. Put another way, conceptions of the power of the weak state should include its positive external image.

The image of a state is an important factor in how we treat it and its citizens. Most black persons traveling with Nigerian passports to Europe or the United States will agree.[2] Indeed, countries that have little else in clout may rely just on their good image to seek to influence the behavior of actors in the much more powerful countries. Yet, in the fields of development and African studies, literature exploring the possible correlation between Africa's negative image in developed countries and the continent's development pace is scarce.

It is not that there is no scholarly interest in Africa's external image. In fact, there is a good body of literature on Africa's image in the West in general, and on its media image in the United States in particular. A journal of the African Studies Association, *ISSUE*, even devoted an entire issue (1994, winter/spring) to this subject. What is missing is the examination of negative external image as an independent variable (Onwudiwe, 1996). This omission is curious, given that the relation between states "is dominated by bargaining power" (Rummel, 1979, p. 4, cited in Ray, 1995, p. 196). Bargaining between two (or more) states, be it for trade, aid, investments, or treaties, is always latently aided by one or more attributes of power. For weak states, a good image is one such latent dimension of power. As Walter Lippman noted, our behavior toward others is informed by the stereotypes in our head: "[W]e all notice a trait which marks a well-known type and fill in the rest of the picture by means of the stereotypes we carry in our heads" (Michalak, 1992, p. 4). In a bargaining situation between a weak and a powerful state, the weak state is better off (i.e., in a stronger bargaining position) if its stereotype (image) in the mind of the powerful state is positive.

Let us refer back to our United States/Ghana example. Ghana's good image can be equivalent to a person or state with a good credit rating in the financial markets. Ghana is rewarded with good lending terms. This objective market rule may not apply to the United States in its relations with Ghana. The United States does not have to be beautiful because it is powerful. In all the traditional measures of tangible power, the United States scores much higher than Ghana. Put another way, the United States controls the financial market in our analogy, which is more important than good credit. The weak state has no choice but to be impeccably clean if it desires approval and good lending terms. Ghana's appeal to those U.S. officials whose decisions in the name of the U.S. state determine how much of its goals Ghana achieves in the United States can be a measure of Ghana's power in its relation with the United States.[3]

Investors respond negatively to grim images of discord, instability, inefficiency and other types of one-dimensional representation of Africa. Even within countries, the negative image of inner cities does not encourage domestic investments here. In the United States, for example, investors are frequently encouraged through tax breaks and other forms of financial compensation (called enterprise zones) to invest their money and time in areas with a negative image caused by perceptions of the prevalence of violence and drugs. These compensations are designed to lower the perceived higher risk of doing business in areas with a bad image. Inner-city lenders usually charge higher interest rates and even supermarkets located there charge higher prices than their counterparts in more affluent neighborhoods with a more pleasant image.

Of course, banks charge lower interest rates to their best customers for reasons that go beyond impressions—credit record, employment, income, and the like. However, image, per se, is a factor. Elsewhere, we have shown how different lending conditions affecting African and Asian countries of approximate wealth standings may be attributed to differing levels of negative external images (Onwudiwe, 1996).

THE IMAGE OF AFRICA IN THE UNITED STATES

What exactly is the image of Africa in the United States today? From the University of Wisconsin, a large elite American university that has contributed a lot to African studies, we have a sample perception of Africa as presented by Professor Jo Ellen Fair's class:

> Each semester when I ask my students, the majority of whom are white and middle class, to describe for me their images and ideas of "Africa" and "Africans," I get the usual litany of stereotypical, negative, and often condescending descriptions. To my students, "Africa" is: "basket case," "jungle covered," "big game, safari," "impoverished," "falling apart," "famine-plagued," "full of war," "AIDS-ridden," "torn by apartheid," "weird," and "black." Moreover, my students describe "Africans" as "tribal," "underdeveloped," "fight[ing] all the time," "brutal," "savage," "exotic," "sexually active," "backward," "primitive." (Fair, 1993, p. 5)

Anyone who believes, as we do, that a do-it-yourself, bootstrap development model for Africa cannot work in an economically interdependent world, and that Africa's economic imperatives require a broad relationship with developed western economies, must be concerned with this idea of Africa held by future leaders of the world's strongest economy. As business people and executives, would they invest their money in a place that is a "basket case," "falling apart," and "full of war"? Moreover, will the cost of doing business in such a place relative to other developing regions be conceived as much higher? The risk factor is certainly higher for businessmen who will have to set up shop in a "primitive" place where they will have to put up with "savages" who are "brutal" and "fighting all the time," and probably rapists, since they are "black" and "sexually active." Africa is simply a risky place for investment. It seems evident that this kind of negative image of Africa in a powerful country can retard the continent's development. If this is the image of Africa in the classrooms of liberal Wisconsin, imagine what students at more conservative campuses think of the continent and its people.

We assume that the development of the African continent depends, in part, on learning from a variety of more developed countries, ranging from the United States, Japan, and countries of the European community to Taiwan, Singapore, Hong Kong, Brazil, and South Korea. Apprenticeship training would rapidly increase the level of skills in Africa's developing economy. Many opportunities for this training, however, are dependent on foreign direct investment. Development experts agree that central to the many mechanisms involved in development is the transformation of societies from low to high income through higher productivity of the labor force. To the extent that foreign trade and investment help in transferring some of the skills that improve productivity of the labor force, obstacles to such trade and investment can be detrimental to Africa's development.

This book provides reasons the bleak view of sub-Saharan Africa should be re-examined and balanced. The various chapters provide quantitative and analytical evidence of Africa's successes despite enormous obstacles. The obstacles are both internal and external. Internally, they may involve the challenge of managing change, for the transformation to modernity involves painful cultural adaptations.

Minabere Ibelema, in Chapter 2, measures Africa's progress by the yardstick of successful adaptation to change while conserving cultural values. Even as they discarded unproductive traditions, African governments maintained and encouraged the growth of aspects of African culture to the level of international prominence, as shown by F. Abiola Irele in Chapter 3. But neither cultural achievements nor the scientific advances detailed in Chapter 10 by Richard W. Hull have been able to help the image of Africa in developed countries.

THE BURDEN OF AFRICAN STATES

For the most part, the pace at which post-colonial African countries are adapting to modernity depends on the condition under which colonialism left them. The patterns they had to follow in economics and in politics were laid down by outgoing colonial powers. As we shall argue, those who sought nationalist directions were instantly punished at the expense of Africa's development. Therefore, the undemocratic colonial legacy, complete with its built-in seeds of crises, including pent up demand born of the conspicuous consumption of colonial officials, wide gaps in black/white incomes, an unhealthy sense of ethnic chauvinism resulting from colonial policies of divide and rule, and particularly political boundaries carried over from the armchair partition of the continent in Berlin (1884–1885), became parts of what Emerson (1967) termed the "burdens of African States" in his presidential address at a meeting of the African Studies Association in 1966. We shall discuss briefly only two areas of these heavy burdens (the economic and the political) with which new African leaders have had to contend, since the natural and man-made constraints inherited by the new states of Africa were relatively recently summarized in Chazan, Martinez, Ravenhill and Rothchild (1992), among other works.

Economic and Political Legacies

Boahen (1987) contains clear and judicious statements on the benefits and limitations of the colonial impact on Africa. How colonialism joined Africa with the world capitalist economy is noted here because that integration involved first and foremost, the penetration of African economies by the noncompetitive capitalism of colonialism. The blueprint for Africa's post-colonial dependency was drawn not so much by colonial economic domination, but, as Ake (1981) noted, by the process of penetration of Africa by colonial capitalism. The colonial economies inherited by the new African states were not shining examples of free market capitalism, complete with sufficient industrial foundations to warrant hopes for future growth and prosperity. Instead, they were characterized by severe monopolistic tendencies, and disarticulation and incoherence in almost all the most important sectors including transportation, manufacturing, and commodity export. As Ake argued, under a disarticulated condition, there is no complementarity between the different sectors of the economy. Under an incoherent condition, the constituent regions and/or sectors of the economy are neither complementary nor reciprocal. As a result, at independence Africa's economies lacked forward and backward linkages in production. In brief, colonialism left neither an example of competitive markets nor enduring foundations for economic development.

In politics, post-colonial Africa was faced with a totally different and higher set of norms under which to operate that was totally different and higher than that experienced under the colonial predecessors. Self-rule meant the end of colonial forced labor. The so-called boom in commodity exports during colonial times was, after all, made possible by forced African labor. Many of the millions of African victims of the system of forced labor suffered and died. For example, Ake notes that 25 percent of the forced laborers who worked for the French on the Brazzaville-Point-Noire railway, which began in 1921, died annually. Furthermore, forced laborers in the Gold Coast (Ghana) working as carriers for

England averaged about 400 miles per month until most became incapacitated, their soles largely worn through. As a corrective action, the British officials tarred the carriers' feet (Ake, 1981, pp. 68–69).

In either case, there was no International Labor Organization to intervene, no human rights group to protest, and no foreign government to threaten sanctions. So the unrestrained inhumanity of the colonial system engendered high productivity and economic success. Yet, those who unfavorably compare post-colonial and colonial African economies often forget these significant facts.

After independence, African governments could not generate the level of revenue required to support the needs of what had to be both developing and welfare states. They were not going to run the system with as much insensitivity to people's pains and lives as the Europeans did; and even if they had tried that mode of heartless accumulation, the international system of which they were now a part as sovereign states would have rightly tried to serve as a wedge.

Unlike European states during their own periods of underdevelopment, for example, the new African statesmen did not choose the mean comfort of lingering "in the relative simplicity of laissez fair, night-watchman states" (Emerson, 1967, p. 6). This is one reason the poor man must be better off in Africa today than he was in Europe in the past. Ibelema provides support for this view in his discussion of African culture in Chapter 2. Although masked by relative material poverty due in part to the subsequent reductions in export income in the new post-colonial non-forced-labor economies of Africa, citizens have been much better off since independence than before.

Still, no matter the reason, a reduction in the revenues of the new African states has important implications for African governance. In particular, it fostered neocolonialism by sowing fresh seeds of dependency of the new African states on their former colonial rulers. This immediate dependency (soon after independence), born of new, more humane (albeit imperfect) behavior of government toward subjects, which was absent in colonial times, required a continuing and special relationship with former colonial powers. The irony is that for the new African states to afford to be humane in their relationship with their subjects, they needed economic aid from former colonial powers from whose inhumane policies they were liberating their African subjects in the first place.

The aid came with tight conditions, one of which was partner concentration in international trade and business. This worsened Africa's terms of trade by constraining its choice of sources for international trade and investments. The resultant, almost monopolistic presence in African economies, guaranteed influence in domestic policies of the new African states by the former colonial powers. This, in part, is the necessary beginning of the phenomenon known as neocolonialism, or indirect political and economic influencing of (in this case) independent African countries.

Second, the colonial economic system and trade undermined the growth of Africa's middle class in many ways, among them, through the system of forced labor and slave wages (already mentioned); the monopolization of trade through the harassment and destruction of creative and ambitious African middlemen at a pace reminiscent of pre-colonial times; and the limited technical training and education of Africans, which reinforced the division of labor in which Africa's role was to supply abundant but largely unskilled labor. The political result is that come independence, Africa had limited manpower to run the inherited institutions of government and to modernize.

Despite limited revenues, huge investments had to be made in education by new African states. It was not a matter of right or wrong economic ideology or policy, it was a clear case of national survival. Some African states instituted free or highly subsidized education. This necessary proclivity toward a social welfare state led to an increasing role of the new African states in the private economy, a condition which demanded centralization of political power. These and other pressures toward concentration and centralization of power were not created by "corrupt" African leaders in a fit of some devious "prebendal and patrimonial" mood itching to reach the hellish condition of "political decay."[4] Pressures simply resulted, and are still resulting, from the inherited conditions of colonialism.

We admit that there is ample abuse of power and mismanagement by independent African governments to merit critical attention, but the attention has often yielded distorting analysis. According to Mamadani (1996, p. 12), the main inclination of the negative approaches to the study of Africa "is to lift a phenomenon out of context and process. The result is a history by analogy." We concur.

Democracy is not alien to Africans. As Njoku E. Awa argues in Chapter 6, democracy is not a European export to Africa either. In modern Africa, its existence or its death has not been solely an internal choice. This is a fact of the Cold War more than it is a measure of African leadership.

In summary, the great need of Africans at independence called for more government revenues, which explained the initial centralization and expansion in the size of government. At independence, Africa did not have a middle class strong enough to block the centralization or to limit the size of the new African states. However, the statist school in a classic blame-the-victim state of mind decided to put an emphasis on the abuses of African leaders rather than on the historical and socioeconomic realities of Africa's peculiar condition at self-rule. While excessive focus on negative outgrowths of statism (abuses of power, corruption, and the like) make interesting thesis, as in Ayittey (1992), they miss the context of Africa's pressing goals as identified, for example, by Kenya's Tom Mboya in 1963:

> The simple peasant may think of Uhuru in terms of farm credits, more food, schools for his children. The office clerk may see it as meaning promotion to an executive job. The apprentice may interpret it as a chance to qualify as a technician, the school boy as a chance for a scholarship overseas, the sick person as a provision of better hospital facilities, the aged worker as the hope of pensions and security in old age. (Mboya, 1980, p. 478)

These concerns are what self-rule or freedom from colonialism (Uhuru) meant for the majority of Africans. They are what decolonization was expected to bring. They constitute the meaning of self-rule: the end of hunger, the quest for higher education, for status, for better health care. For millions of Africans, self-rule was the realization of the hope that things would become better for the poor, or as Fanon (1970, p. 490) put it a long time ago, a time when "the last shall be first and the first last." Of course, good things include good governance, but that is just one of many values.

We do not suppose that Mboya's inventory of expectations has changed much since the early 1960s, nor should we assume that Africans are waiting for some free and fair election to bring forth freedom of the individual. Peter P. Ekeh shows in Chapter 5 how, in spite of the behavior of any given African state, the domain of kinship acts to safeguard individual liberties in the continent. Along with the continuing (albeit slow) growth and expansion of the ownership of the private economy by the middle class, it should be expected that the role of government in African countries will incrementally recede, as will opportunities for abuse of political power. Slowly but steadily, post-colonial African governments have been successfully building the infrastructure relevant for the full realization of this hope in various sectors. Health is one of those sectors, as Emmanuel U. Nnadozie argues in Chapter 9; transportation/communications is another, as Folu Folarin Ogundimu and Okechukwu C. Iheduru demonstrate in Chapter 8. These developments are cause for hope, not despair.

The Legacy of the Cold War

Following on the heels of the World War II was the East-West rivalry for influence over the new African states, which ended with the collapse of the Soviet Union in 1991. The hope in Africa since then has been for the post-Cold War system to bring forth a more pleasant international order under which the development of Africa will not be held hostage. In Africa, the Cold War had encouraged surrogate bloody wars, introduced military coups, and propped up local military and civilian dictatorships. African leaders had little time and independence of action to formulate and to implement development policies as they saw fit. It is, therefore, our belief that one key to understanding the performance of self-rule in Africa lies in analyzing Africa's place in the bipolar international environment of the Cold War era.

The most disastrous effect of the Cold War on Africa was in raising the political cost of patriotism. Nationalist politicians were out of favor; internationalists carried the day. Africa's domestic policy issues under pragmatic internationalists were increasingly subordinated to priority issues of bipolar security preoccupations. In the process, emerging African voices against structural inequities in the international political economy were frustrated, denationalized, overthrown, or killed. At the international level, millions of dollars that could have been used for the development of Africa's new post-colonial states were squandered on military competition as Africa became—to borrow a phrase from Foltz and Bienen (1985, p. 8)—one of the "surrogate terrains," where superpower rivalries were played out. Though Foltz and Bienen were actually describing Africa in the great power strategy before self-rule, their description perfectly fits the era of post-colonialism. For after independence, Africa was still one of the international scorecards for tallying the gains and losses of the two macho ideologies of the Cold War: capitalism and communism.

For some African patriots, the price was more than political. It meant death for Patrice Lumumba of the Congo; disgrace, exile, and death for Kwame Nkrumah of Ghana; ostracization for Sekur Toure of Guinea; harassment and long-term jail for Nelson Mandela; torture and death for Steve Biko of South Africa; and torture, imprisonment, and death for countless and nameless others. These severe, rapid, and open punishments meted out to patriotic and nationalist Africans served as deterrence to would-be nationalists, those real leaders who would dare to put Africa before the demands of the Cold War camp of their respective states.

This situation discouraged creative pro-African policies by most African leaders. Given the high cost of such policies, was it not rational for the decision makers to go along with policies that met the needs of the foreign leaders of their global ideological leaders? In anglophone, francophone, lusophone Africa, the United Kingdom, France, and Portugal, respectively, were ever-present influences that counterbalanced pro-nationalist proclivities of non-"moderate" and "unreasonable" African politicians. This way, some desirable people were effectively excluded from positions of power, and in their places were some willing and unwilling partners in the neocolonial arrangement. These people were the realists, the pragmatists, the moderates, the "reasonables" of which Mobutu Seso Seko was an unmistakable example.

After many years, most members of independent Africa's political and economic elite began to get used to the new condition for surviving in the new world order whereby economic and especially political power in the domestic environment depended on conforming with the norms of bipolar politics. Soon, most modern and cosmopolitan Africans, leaders, and would-be leaders alike, discovered that these norms of the new order of things had both luxurious rewards and deadly punishments. Thus, unlike their counterparts in nondependent states, most African leaders were (and still are) torn between the desire to do good by Africa and the need to do good by external powers. These leaders served (and are still serving) double constituencies.

DOUBLE CONSTITUENCIES, DOUBLE LOYALTIES

A miracle of post-colonial Africa is that despite the subsequent inner tension, its leaders managed, to some extent, to meet the needs of these double constituencies: development and pent up demands of citizens of the type Mboya spoke to, and policy conformation demands of Cold War powers. It is no wonder that the founding fathers of Africa's political independence chose the ideology of pragmatism over any other. They simply did not have the luxury for emotional attachment to either of the Cold War ideologies that made their lives so difficult. One of Nigeria's founding fathers, the late Chief Obafemi Awolowo clearly captured this necessary ideology in his autobiography (Awolowo, 1960, p. 283): "The emphasis as far as I am concerned, has always been on the words 'expedient and effective.'" In powerful countries, governance is relatively simple. Policy is shaped by people who, for the most part, derive their power from a single constituency—their citizens, who in turn give them support through taxes, votes, and lives. Africa's post-colonial leaders had two constituencies or sources of power, and therefore, double loyalties. One half of the loyalty was to foreign powers who supported them with financial, political and military aids, and protection, and who in return expected (and often got) policy compliance, even when such policies were not in Africa's best interest.

This context of Africa's self-rule has not significantly influenced statist critique of the performance of post-colonial African government. The tendency has been to subsume it under image-distorting categories of corruption, misrule, and political decay. No matter the relevance of these categories, they are only one side of a more variegated story. This book tells the other story—the story of African progress—and the individual chapters provide inventories of that progress.

The chapters are grouped by subtopics into six parts, including this introduction, which constitutes Part I. Part II examines culture, identity, and development. The placement and number of chapters (three) in this section are intended to suggest the centrality of culture in societal well-being and development, as is argued in the first chapter in this section, Chapter 2.

In Chapter 2, Ibelema notes that the process of development entails some transformation of a people's way of thinking and that the challenge to post-colonial African states has been to reconcile change with tradition, and to determine what tradition is worth conserving and what should be discarded. The chapter entitled "Tradition and Modernity: The Triumph of African Culture" examines how successfully African countries have met this challenge. Focusing on changes in attitude toward innovation, which entail both qualities of openness to new ideas and a critical stance, Ibelema notes that innovation could introduce dislocative effects beyond its agenda, especially in the absence of an orderly transition or integration into the traditional order. Noting also the challenge of measuring such transformations, Ibelema innovatively uses changes in attitudes indirectly gauged with voting behavior, political alignment, economic activities, health practices, and social relationships to bypass the measurement problems. He celebrates the resilience of the core elements of African culture such as communalism even as changes in attitude admit new ways of doing things. The chapter employs a variety of analytical approaches, including participant observations, anecdotes, and direct experience to demonstrate African transformations in traditions, education, gender values, and health care, among other spheres.

In Chapter 3, F. Abiola Irele, a leading literary critic, notes that although the contemporary situation in Africa is generally perceived as bleak, the attribute of creativity continues to manifest itself as vigorously as ever in various spheres of cultural expression, such as the cinema, music, visual arts, and literature. His chapter, "Artistic Creation in Post-Independence Africa," provides a cogent inventory and analysis of achievements in creative work. In popular music, for example, *juju* from Nigeria, *soukous* from Zaire, and *makossa* from Cameroon are, like jazz, internationally recognized musical styles. In literature, the establishment of local publishing houses has helped to promote the work of a younger generation of writers in a way that is now promoting a veritable renaissance in some African countries. Irele's analysis crystallizes the continuity and vitality of Africa's cultural expressions.

In Chapter 4, Tracy D. Snipe provides a case study of cultural policy and its place in politics and national identity. In the chapter, "Cultural Politics in Post-Independence Senegal," Snipe examines one African government's programmatic activities at cultural conservation, celebration, and innovation. The choice of the government of Leopold Senghor, the proponent of the philosophy of negritude, could not have been more apt.

Part III examines governance and the political order. The distinguished political scientist Peter P. Ekeh begins this section with Chapter 5, entitled "Kinship and Civil Society in Post-Colonial Africa," an examination of the historical contributions of kinship and related institutions to the general welfare of Africans. The chapter reviews the theory, functions, and structures of civil society in a way that puts the contemporary application of the construct to Africa in proper historical perspective. The longevity of kinship in Africa is itself an important achievement given the ameliorating role it plays in the relationship between Africans and African states. While analyzing this important function, Ekeh reveals an important difference between the freedom of the individual due to kinship and freedom in liberal democracy even as the former undergoes qualitative changes in the post-colonial period when African states are forced to recognize the autonomy of their primordial public domain.

In Chapter 6, Njoku E. Awa argues that participatory democracy in Africa is not a European import. The chapter reviews the ways by which the methods of participatory decision among the Igbo-speaking people of southeastern Nigeria have been gradually introduced at local government levels. Awa notes that the spirit of participation in Africa is not at all moribund. "Rather, it has been repressed by a history of conquest." The chapter analyzes some important foundations for the democratic process in Africa.

Part IV examines economic and infrastructural developments. In Chapter 7, H. Sylvain Boko examines Africa's successes in the very sphere that Africa is said to have failed abysmally, economic development, as measured by the traditional indices. By providing three in-depth case studies—Benin, Botswana, and Mauritius—Boko demonstrates that several economic success stories have been overshadowed by the accounts of regression and gloom. Boko shows also that even when the African picture is taken in totality, there have been post-colonial periods of economic success.

In Chapter 8, Folu Folarin Ogundimu and Okechukwu C. Iheduru review achievements in transportation and communications. Using data gathered on selected African states, they show how transformations in communications and transportation since independence have increased the ability of Africans to gain access to modernity. The modernization process included the dramatic increases in the diffusion of electronic communications receivers (radio and television) and elevated contacts with the outside world through new and improved roadways and waterways.

Part V examines health and agricultural innovations. In Chapter 9, the economist Emmanuel Nnadozie examines the performance and improvement in health policy and health care delivery by African governments during the post-colonial period and shows that there have been improvements since the 1960s. Nnadozie uses country time-series performance analysis to show quantitative improvements in basic health indicators including related manpower, hospitals, mortality rates, and health expenditure.

The field of science is another area of quiet but steady post-colonial progress. This is significant because it indicates the creative role of modern science in post-colonial Africa in the battle against hunger and disease. The role of modern science in the development process cannot be overemphasized.

In a fascinating analysis full of empirical evidence, historian Richard W. Hull calls attention to the significant progress that has been made in Africa in the fields of bio-genetic engineering and preventative medicine. In Chapter 10, "Biotechnology, Food Production, and African Advancement," Hull summarizes the current research and successes of Africa's major institutes of tropical agriculture and medicine. Theories of underdevelopment are often stretched until they overshadow or minimize internal contributions of Africans and friends of Africa toward African progress. Paulin J. Hountondji's notion of extraverted scientific research whereby research agenda of African scientists in the periphery are primarily defined by the needs of their counterparts in the core countries is one example of this observation (Hountondji, 1992). However, Hull's impressive enumeration of liberating scientific achievements in eradicating such diseases as malaria, sleeping sickness, and river blindness and in the development of new crops and the hybridization of traditional staples (including sorghum and cassava) puts into question some aspects of Hountondji's epistemological dependency and "consumer of science" hypothesis. In addition, the chapter evaluates the adoption of low technology, cost effective methods of soil and water conservation, and the dissemination of the resultant knowledge to rural areas.

The book concludes with Part VI, in which the lone chapter examines the future of Africa. In Chapter 11, entitled "Afrenaissance: Struggles of Hope in Post-Colonial Africa," the eminent scholar Ali A. Mazrui defines Afrenaissance as "living evidence of Africa's self renewal, living testimony that the human spirit is reasserting itself among the African people." The chapter encapsulates the spirit of this book and projects it into the future. It challenges the selective judgment of the international press who are quick to report abuses of African governments even as they ignore the heroic counteractions of ordinary Africans. When it comes to Africa, dictators and villains get more coverage and pro-democracy riots are reported as political instability. In Europe, on the other hand, the defiance of Solidarity under Lech Walesa was reported as heroic and he was subsequently rewarded with the Nobel Prize. This is the specific character of the double standards in international reporting. But more than these valuable insights, the chapter is vintage Mazrui, with its stimulating but potentially controversial proposals on conflict prevention, containment, and resolution in Africa.

We referred earlier to the distinguished philosopher V. Y. Mudimbe, whose widely acclaimed work *The Invention of Africa* inspired this project. We must add that ultimately Africans will have to reinvent themselves or reverse the invention that Mudimbe aptly discussed. The process has long started.

NOTES

1. A limited but good reference to anti-African scholarship can be found in Thompson (1985).

2. The special suspicion and humiliating search that Africans (especially Nigerians) attract at Heathrow in London, JFK in New York, and other major airports in Europe and the United States is well known to this writer and all other African travelers. Even Wole Soyinka, a Nobel laureate in literature, was subjected to similar treatment at JFK. Negative international image is an equal opportunity humiliator of all Africans.

3. Except when national interest intervenes. Even then, the ability of the weak state to maximize its gains in a given situation will still be relatively limited. We believe this to be also true in the private sector.

4. A good and concise review of various trends in studies of African politics including these types of negative approaches is in the first chapter of Chazan et al. (1992).

REFERENCES

Ake, C. (1981). *A political economy of Africa.* New York: Longman.

Awolowo, O. (1960). *Awo: The autobiography of Chief Obafemi Awolowo.* Cambridge, UK: Cambridge University Press.

Ayittey, G. B. N. (1992). *Africa betrayed.* New York: St. Martin's.

Boahen, A. A. (1987). *African perspectives on colonialism.* Baltimore, MD: Johns Hopkins University Press.

Chazan, N., Martinez, R., Ravenhill, J., & Rothchild, D. (Eds.). (1992). *Politics and society in contemporary Africa.* Boulder, CO: Lynee Rienner Publishing.

Curtin, P. D. (1995, March 3). Ghettoizing African history. *Chronicle of Higher Education*, p. A44.

Emerson, R. (1967). African states and the burdens they bear (presidential address, African Studies Association, Bloomington, IN). *African Studies Bulletin, 10*(1), 3–13.

Fair, E. F. (1993). War, famine, and poverty: Race in the construction of Africa's media image. *Journal of Communication Inquiry, 17*(2), 5–22.

Fanon, F. (1970). Concerning violence. In E. Kedourie (Ed.), *Nationalism in Asia and Africa* (pp. 488–539). Cleveland, OH: The World Publishing Press.

Foltz, W. J. and Bienen, H. S. (1985). *Arms and the African: Military influence on Africa's international relations.* New Haven and London: Yale University Press.

Hammond, D. & Jablow, A. (1970). *The Africa that never was: From centuries of British wishing about Africa.* Prospect Heights, IL: Waveland Press, Inc.

Hawk, B. G. (Ed.). (1992). *Africa's media image.* New York: Praeger Publishers.

Herrnstein, R. J. & Murray, C. A. (1994). *The bell curve: Intelligence and class structure in American life.* New York: Free Press.

Hountondji, P. J. (1992). Recapturing in V. Y. Mudimbe (Ed.), *The surreptitious speech: Presence Africaine and the politics of otherness, 1947–1948.* Chicago: The University of Chicago Press.

Ibelema, M. & Onwudiwe, E. (1994). *Today* in Africa. *Issue: A Journal of Opinion, 22*(1), 10–12.

Irele, A. (1990). *The African experience in literature and ideology*. Bloomington: Indiana University Press.

Johnson, P. (1993, April 18). Colonialism's back—and not a moment too soon. *New York Times Magazine*, pp. 43–44.

Johnson, P. (1992, Dec. 14). Wanted: A new imperialism. *National Review*, pp. 28–32, 34.

Kaplan, R. D. (1994, February). The coming anarchy. *The Atlantic Monthly*, pp. 44–76.

Mamadani, M. (1996). *Citizen and subjects: Contemporary Africa and the legacy of late colonialism.* Princeton, NJ: Princeton University Press.

Mboya, T. (1970). The mass movement. In E. Kedourie (Ed.), *Nationalism in Asia and Africa* (pp. 477–487). Cleveland, OH: World Publishing.

Michalak, S., Jr. (1992). *Competing conceptions of American foreign policy: Worldviews in conflict.* New York: Harper Collins Publishers.

Morganthau, H. (1967). *Politics among nations*. New York: Knopf.

Mudimbe, V. Y. (1988). *The invention of Africa.* Chicago: Bloomington: Indiana University Press.

Onwudiwe, E. (1996). Image and development: An exploratory discussion. *Journal of African Policy Studies, 1*(3), 85–97.

Ray, J. L. (1995). *Global politics.* Princeton, NJ: Houghton Mifflin.

Rummel, R. J. (1969). Indications of cross-national and international patterns. *American Political Science Review, 63*, 127–147.

Thompson, L. (1985). *The political mythology of apartheid.* New Haven: Yale University Press.

Part II

CULTURE, IDENTITY, AND DEVELOPMENT

2

Tradition and Modernity: The Triumph of African Culture

Minabere Ibelema

While a universally accepted definition of national development has eluded scholars and policy makers alike, there is the consensus that it entails some transformation. To capitalist economic theorists such as Adam Smith, Max Weber, and Walt W. Rostow, development comes in the form of increased capital accumulation and investment; to Marxists, it derives from the transfer of economic leverage from the bourgeois to the proletariat; and to adherents of the differentiation theory (largely sociologists and political scientists), the emphasis is on social changes that facilitate fluidity in individual's roles and the differentiation of functions (Teheranian, 1977). Whatever the form of change envisaged by a given theory, it has to be preceded or at least accompanied by a corresponding transformation in the people's way of thinking.

Therefore, innovation is typically a challenge to tradition (including cherished and beneficial tradition) and may introduce dislocative effects beyond its agenda. For these and other reasons, Daniel Lerner's (1958) seminal work *The Passing of a Traditional Society* has been criticized for suggesting that development requires a sweeping abandonment of traditional values through the agency of the mass media. While Lerner may have indeed overstated the case for transformation, his basic tenet remains valid. The challenge for developing countries is to reconcile change with tradition: to determine what tradition is worth conserving and which should be discarded, what innovation should be adopted and which should be avoided. This entails, on the one hand, an openness to new ideas and, on the other hand, a critical, conservative stance.

Parker and Mohammadi (1977) have noted that a major flaw in the classical paradigm of national development is its emphasis on tangible indices such as Gross National Product, per capita income, and infrastructure. Yet, "[a]lthough the improved quality of life that economic development brings is usually the major goal of development, economic indicators are imperfect measures of the quality of life for which they are surrogates" (Parker and Mohammadi, 1977, p. 168). Such criticism has led to an increasingly more expansive definition of development. In the years following World War II, it "became synonymous with growth, modernization, change, democracy, productivity, industrialization, and a score of related Western historical and evolutionary changes" (Mowlana & Wilson, 1990, p. 5). More recently the United Nations Development Programme (UNDP) has further refined development to include what it calls human development indices, including communications, health care resources and factors, education and literacy, gender equity, and social order (United Nations Development Programme, 1999). Although Mowlana and Wilson have criticized the broadened concept as lacking in "the rigorous epistemological, conceptual, and at times, methodological power" of the orthodox concept in explaining the transformation of societies, it may be said that by that very quality, the expansive conception of development more closely approximates the complex reality it describes.

In fact, even the UNDP's more inclusive index hardly reaches the essence of societal transformation. We argue that it is not enough to note the physician-population ratio and the availability of health clinics. It is also important to acknowledge the increase in the proportion of people who want to visit hospitals and consult with doctors. It is not enough to itemize the statistics on educational structures and pupil-population ratio. It is also important to examine changes in attitude regarding the importance of formal education and who may or may not go to school.

Of even greater importance is society's adjustment to the cultural tension inherent in such transformations. As Eisenstadt (1976, p. 41) writes, "These processes of growing differentiation, social mobilization, and breakdown of traditionality—unparalleled in the history of human societies—create within the societies in which they occur certain basic problems of regulating the necessarily increasing conflicts that develop among the newly emerging groups; of integrating these groups within some common institutional framework; of developing some foci of collective national identity in which tradition, modernity, and change are to some extent combined." This chapter examines such intangible changes and adjustments in Africa and notes their significance. We contend that one of the most significant achievements of post-colonial Africa has been the simultaneous transformation and conservation of cultures despite the dislocative effects of colonialism and other nascent hegemonic and modernizing pressures on African culture.

First, we examine the concept of culture, then we discuss the historical and contemporary stresses on African culture, including the impact of colonialism. Then we demonstrate that African culture has withstood the stresses and evolved. The ultimate point of this chapter is that one cannot adequately gauge

Africa's post-colonial advancement without examining its success in cultural adaptation and conservancy.

THE CONCEPT OF CULTURE

In its broadest sense, culture is the totality of a people's way of life. In a narrower sense, it refers to specific, perhaps unique, sets of values, such as Weber's (1958) notion of the "Protestant ethic."

Culture has also been defined in functionalist and instrumental terms. Clyde Kluckhohn (1962, p. 54), for instance, has defined culture as "all those historically created designs for living, explicit and implicit, rational, irrational, and nonrational, that exist at any given time as potential guides for the behavior of men." Malinowski (1944) has gone even further to posit that all cultural elements have their origins in organic, that is, primary, needs, the satisfaction of which gives rise to secondary or social necessities. "The understanding of any cultural element must imply, among other things, the statement of its relationship, instrumental or direct, to the satisfaction of essential needs, whether these be basic, that is biological, or derived, that is cultural" (Malinowski, 1944, p. 142). Thus, Malinowski explains a range of societal activities and beliefs, ranging from eating habits to superstition.

Whether culture is conceived of in its broadest or narrowest sense, it is invariably a social stabilizer and compass. It is the way a given people have grown to adapt to their environment, to rationalize their existence, to find order and, perhaps, even happiness. Even if one does not completely accept Malinowski's notion of instrumentality of culture, there is little doubt that changes enforced on a culture without the due processes of rationalization would be damaging.

Of course, change or modernization is an inherent aspect of culture. In fact, an insular culture risks regression and irrelevance. In an innovative culture, on the other hand, there is a growing "capacity to empathize or an ability to see the world from another man's perspective, as well as growing capacity for self-doubt, self-examination, and self-correction—in short, the growth of an open personality" (Teheranian, 1977, p. 32). Cultures, like individuals, have to change—by their own logic—as the society adjusts its customs to the necessities that engendered them.

As diffusion studies have demonstrated, how change is introduced makes considerable difference in how it is accepted and its dislocative impact on society (Rogers and Shoemaker, 1971).[1] As Ngugi (1993, pp. xv–xvi) writes, in making a biological analogy:

> Under 'normal' circumstances, a given society is able to absorb whatever it borrows from other contacts, digest it and make it its own. Under conditions of external domination, conquest for instance, the changes are not as a result of the working out of the conflicts and tensions within, and do not arise out of the organic development of that society, but are forced upon it externally.
>
> This may result in the society becoming deformed, changing course altogether or even dying out.

The decline of the Egyptian civilization following the Roman invasion readily illustrates this phenomenon (Diop, 1990). In recent history, African cultures have had to cope with comparable stresses, the most dislocative of which emanated from colonization. In addition, African cultures have had to cope with the effects of modernizing economies and lifestyles and the pervasiveness of Western mass media. These will be summarized forthwith, followed by a discussion of the processes of adaptation.

First, it should be noted that the notion of foreign influences on African culture is problematic given that what is now seen as foreign may have been African-inspired during earlier ages, as Diop (1990) among others have demonstrated. The notion is applied in this analysis nonetheless to describe values, means, and technologies that depart from the norm of African life during the periods under analysis and could not have resulted from the internal process of cultural evolution. Also, given the academic debate over whether we should speak of an African culture or African cultures (Asante, 1990; Soyinka, 1990), it should be noted that both notions are used in this chapter as deemed appropriate. This should not be read as contradictions or inconsistencies. The phrase "African culture" is used here to refer to commonalties in African life, some of which are referred to or implied throughout the essay. On the other hand, there are elements of uniqueness and even differences among Africa's various peoples. Therefore, the phrase "African cultures" is used when the intent is to emphasize that diversity.

COLONIALISM AND AFRICAN CULTURE

Swedish sociologist Johan Galtung has stated that "Imperialism is a structure; it is not a collection of wicked people, although it may make ample use of such people" (Galtung, 1980, p. 113). To this apt description may be added that colonialism is not merely the physical occupation or economic usurpation of a place or people, it is also a process of psychological transformation which brings about the re-ordering of self and society in relation to others. The colonized are induced to believe that their conquerors are superior not just for the means of coercion at their disposal, but also in their natural abilities and even culture. The African experience in this regard was made acute by the widely shared view, in the words of Arnold J. Toynbee, that "The black races alone have not contributed positively to any civilization as yet" (quoted approvingly by Isherwood, 1966, p. 29).[2]

Some historians have claimed that the goal of early European inquest into Africa was not to subjugate Africans but to bring civilization to Africans. MacMillan (1949, p. 9) writes, for instance, that: "In the nineteenth century civilization was thought of as a higher synthesis of the best experience of all the human race. No leader in those days questioned the theoretical right of Africans to equality and our duty to help their 'progress.'" MacMillan further claims that that mission changed only after World War I raised doubts about European civility in the minds of Africans and discredited the ideology of liberalism within the imperial powers. However, neither the policy nor the tactics of early colonizers bear out MacMillan's charitable view.

Even if MacMillan's view is granted, the very notion of a civilizing mission portends cultural crisis for the people to whom the benevolence is directed. Indeed, whether the colonial mission was to dominate or to civilize, the goal was to change an existing order, to "tame," as Chinue Achebe masterly depicted in the classic novel *Things Fall Apart* (Achebe, 1996). But, at least, a civilizing mission would have relied on less coercive measures. Rather, from the outset, colonial administrators were, at best, patronizing in their treatment of Africans and African cultures and, at worst, brutish. Specifically, African cultures were assaulted by statutory and administrative laws, missionary doctrines, and educational policies.

Some cultural expressions were considered a political threat and were banned outright. Such was the fate in Kenya of "such popular, but politically-conscious dances and songs like Kanyegenuri and Muthirigu" (Ngugi, 1993, pp. 44–45). Other cultural practices were treated with contempt or disrespect. Native names of people and places were routinely changed for the convenience of colonialists.[3] The Niger Delta city state of Okoloama (land of the curlews), for instance, was renamed Bonny in apparent tribute to the island's attractiveness. Another community was named Brass probably because the natives often said "bara sin" (Ijaw for hands off) to decline a trade offer.[4] To date a substantial number of the natives of the Niger Delta and other parts of West Africa still go by the last names of Allison, Brown, Douglas, Green, Hart, Johnson, Jumbo, LongJohn, Pepple, Spiff, Wilcox. These were names officially assigned to native leaders for the convenience of colonial administrators and warlords, who would not be encumbered by unfamiliar native names. As required by tradition, the leaders passed the names on to their subjects.

Perhaps the single most important attempt to suppress African culture was by Christian missions. Mudimbe (1988, p. 45) aptly sums up this impact by noting that: "The more carefully one studies the history of missions in Africa, the more difficult it becomes not to identify it with cultural propaganda, patriotic motivations, and commercial interests, since the missions' program is indeed more complex than the simple transmission of the Christian faith." To begin with, the missions insisted that converts forsook African traditions, not only of the religious kind, but of the social and even personal. Thus, converts were not to partake in traditional rituals, dances, songs or to even have native first names. And they were "encouraged to dispose of any artworks which had been used in [native] religious rites" (Fasuyi, 1973, p. 21). The rationale, as Kirby (1994, p. 61) explains, was that the missions believed that, "Proper European civilization *was* Christianity, and the only way to bring about conversion was to establish this cultural framework . . . . What was Christian could, therefore, not abide with what was African."

This doctrine, which lasted up to the period of independence in the late 1950s and early 1960s, gave rise to myriad social peculiarities. To be baptized, for instance, this writer had to be given the "Christian" name Phillip, a name which meaning, "lover of horses," is odd in a place where few had ever seen a horse. Yet, it was deemed more appropriate than Minabere, a name that suggests the family relational climate in which the child was born and which carries an enduring moral message. In the Ibani dialect of the Ijaw language, Minabere literally means "family problem," a warning to meddlers to not fan family conflicts and a reminder to family members to exercise restraint. Even after the family differences in point were resolved, the name continued to serve as a reminder of the sanctity of the extended family. Notwithstanding this very Christian significance, the name was unchristian in the church's dogma. Apparently, it was more Christian to be a lover of horses than to be a lover of the family.

Colonial education complemented religion and politics in the assault on African culture. As Chinweizu (1975) and Mazrui (1978) have noted, much of the content of education had the effect of intensifying adoration of the colonial masters. The first colonial schools were mission schools and accordingly emphasized theology, which, as already noted, extolled western culture. After World War I, colonial governments began to take a more active part in funding education and establishing formal policies. But even then, the emphasis did not change substantially. A document drafted by the Advisory Committee on Native Education in the British Tropical Dependencies and published in 1925 states as follows:

> Since contact with civilization—and even education itself—must necessarily tend to weaken tribal authority and the sanctions of existing beliefs, and in view of the prevailing belief in the supernatural which affects the whole life of the African it is essential that what is good in the old beliefs and sanctions should be strengthened and what is defective should be replaced. The greatest importance must therefore be attached to religious teaching and moral instruction. (Fetter, 1979, p. 131)

The document does not indicate how the good traditions would be strengthened, but as an indication of the drafters' actual concern, it recommends the establishment of local Boy Scouts and Girl Guide movements.

The concerns of German colonial operatives during the period is similarly enlightening. Martin Schlunk, chairman of the German Evangelical Mission Council from 1924 to 1946, was forthright in a policy essay he wrote in 1914 before his appointment to chair the policy-making council:

> It is absurd to think that even German-speaking natives will then be on our side. They will remain children of their culture whether they have learned to speak German or not. On the contrary, those natives who have received an education from us will then become leaders of their people in the struggle against us. The only chance of averting this is to use the schools for character education. The only schools that can do that job satisfactorily are the theological seminaries of the various missions. (Fetter, 1979, p. 127)

The phrase "character education" is quite significant in this context, with its implication that education that results in the rejection of colonialism is morally deficient. Accordingly, much of colonial education was aimed at impressing on the colonized the value of colonialism. A school examination in Togo in 1909, for instance, included this essay question for which 1.5 hours were allotted: "What good things have the Europeans brought us?" To ensure the proper answer to such questions, schools immersed their African pupils in European history and culture. Ngugi (1993, p. 88) writes that "in the school that I went to, Scottish country dances were allowed even as the so-called tribal dances were banned."

Until after independence, the history taught was essentially European history. Textbooks discussed Greek, Roman, Persian, and Egyptian civilizations, the last of which was presumed to be non-negroid. There was little on black African history, which became a part of secondary school curricula close to or after independence. African universities followed similar patterns. As Chinweizu (1975) and Mazrui (1978) have pointed out, many contemporary universities in Africa began either as colonial campuses of European institutions or were established with the collaboration of those institutions. The universities were essentially European in character, content, and orientation. Of course, the positive values of the universities on Africa's advancement cannot be overstated. Still, their contributions to the dislocative pressures on African cultures are just as significant.

The colonial stresses on African cultures exerted a considerable toll, the most insidious of which has to be what is commonly referred to as colonial mentality—the implicit belief that everything European is superior to everything African. That, in turn, gave rise to Africans' classification of themselves as "natives" and "non-natives," the latter referring to those Africans who through education, missionary conversion, or a mere visit overseas could lay claim to immersion in European culture. The term "native" often was used synonymously with primitive. Thus, the natives were the partakers in African traditions, including dances and rituals, to which the non-native Africans could only be admiring spectators at best. Thus, the first significant social impact of the assault on African culture had to be the rise of what the late Obafemi Awolowo, one of Nigeria's first political luminaries, termed "pseudo-Victorian aristocracy" (1960, p. 81). It is an appellation that had both economic and cultural implications.

In summary, the colonial experience caused considerable stresses to African cultures. Although those stresses were relieved by counter-programming after independence, African cultures have continued to experience other pressures, most notably the pressures from modern economies and lifestyles and the pervasiveness of Western mass media.

POST-COLONIAL PRESSURES ON AFRICAN CULTURES

The realities of the modern economy have put considerable stress on African cultures, especially with regard to familial relationships and individual obligations to society. Such a stress is not unique to any one place, but it may be most consequential in Africa where communal values are very strong. Perhaps, nowhere else is the maxim more true that "every man is his brother's keeper." Africa's extended family system is the most widely recognized manifestation of this value. Yet the value has come under severe stress, as African countries transform economically.

African countries typically consist of amalgams of industrial and traditional societies, which means families often find themselves split between two worlds with varying imperatives in lifestyles. As Tehranian (1990, p. 161) has written, "Frequently, the two sectors of the population live in separate quarters, sometimes as if in separate countries and centuries." The physical and orientational distance often limits the opportunities for familial bonding and social immersion in the ancestral home. With modern communication facilities rarely available at both ends, interpersonal communication becomes infrequent and with that an increased likelihood of misunderstanding, distrust, and alienation.

Moreover, traditional economic activities, which used to be sufficient at least for subsistent living, often can no longer suffice. Family members, usually one or two, who are a part of the modern economy are expected to provide for the extended family even when such expectations cannot possibly be met. The limited opportunity for communication only exacerbates matters. The resulting tension has been a constant source of stress in African families, though often muted. The stress has threatened, but not destroyed, the fundamental African value of communalism.

African cultures are also under pressure from international mass media and the dominant Western culture from which they largely emanate. Like most other parts of the world, African countries are in the receiving end of Western-produced informational and cultural products, among them news, film, television programs, and books.

Economic imperatives, elite cultural orientation, and political pressures all contribute to the importation. For instance, in a drive toward modernization, many African countries imported television broadcast facilities just before or soon after independence. With no history of production, they turned to Western exporters for programs, sometimes to satisfy the tastes of a post-colonial elite. In some cases, the political leaders were pressured to import such goods as part of a quid pro quo in their relations with the erstwhile colonial and neocolonial masters (Schiller, 1976).

Research suggests that exposure to such foreign cultural productions have considerable impact both on individuals and society. Psychologist Kenneth Gergen has written of "post-modern consciousness," a phrase he uses to describe the condition of flux in one's sense of self (Gergen, 1991). Gergen argues that when people are bombarded with myriad images, their reality changes incessantly as does their value system. Such people often don't know what to believe in or what to stand for. What Gergen writes in an intranational context has to be doubly valid in the international context, where the recipients have little basis for ascertaining what is plausible and what is not.

Take, for instance, a U.S. film character and narrative that suggests, as they often do, a value orientation. An American would more easily, though not necessarily, assess the extent of realism in what is depicted. He might, at least subconsciously, ask the question, "Is there anyone I know who would act this way?" The answer may well be no. In contrast, someone in another country who is not steeped in American culture is more likely to conclude that what is depicted has a basis in American culture and therefore offers a plausible behavioral option. Thus, a non-American is more susceptible to the suggestive influences of fictional productions, especially if that person sees the United States as the benchmark of modernity and contemporary culture.

These then are the major stresses—historical and contemporary—on African cultures. How African cultures have borne these stresses and, in fact, thrived under them is the subject of the rest of this chapter.

REHABILITATION AND CONSERVATION OF AFRICAN CULTURES

The dislocative effects of the stresses on African culture just summarized were most consequential among the elite and the urbanized. The rural masses were affected too, but not nearly as profoundly. As Lerner (1958) found in his study of the transformation of the Middle East in the 1950s, rural communities are relatively insular. Despite the influence of missionary activity and education, rural African communities remained quite steeped in African cultures even as urban centers reflected rapid transformations. However, the urban-rural distinction has become quite fluid, with the rural population flocking to the urban centers and urban dwellers routinely returning to their native villages either to visit or to stay. Inevitably, what transpires in the urban centers ultimately transforms the rural centers. The rehabilitation of African cultures then is a process primarily among the elite and urban classes and secondarily among the rural masses.

To rehabilitate suggests that the injurious act is completed and the injured awaits healing. With regard to African cultures, the stresses are ongoing. Therefore, rehabilitation has to work in tandem with prevention of further injuries, in this case through conservation and even cultivation. Several African leaders recognized quite early in the colonial era that the preservation of African cultures had to be the foundation of Africa's development. Therefore, the rehabilitation and conservation of African cultures began quite early in the process of political decolonization. The primary goal was to restore and assert a sense of dignity in Africanness.

Accordingly, as Mazrui (1978) has noted, many early political parties began as cultural organizations. These include Nigeria's Egbe Omo Oduduwa and Jami'a, which gave rise, respectively, to the Action Group and Northern People's Congress, both of which were major political parties in pre-independence Nigeria and several years after independence. Similarly, Kenya's "Jomo Kenyatta was a cultural nationalist before he became a nationalist of any other kind." (Mazrui, 1978, p. 58). Post-independence governments continued on this track by promoting traditions and cultures through the mass media, festivals, grants, and regulatory processes.

In Nigeria, for instance, federal expenditures for arts and culture rose substantially in the 1950s, during the period of incremental transfer of colonial power. In 1960–61, the first year of independence, £61,440 were allocated for arts and culture, representing 3.6 times the amount allocated in 1950/51 (Fasuyi, 1973). About a decade after independence, the allocation had increased to £158,450, or about 2.6 times that of 1960/61. These increases were substantially higher than what could be accounted for by inflation or increases in government revenue.

The budgetary allocations were in keeping with the objectives of the Nigerian Arts Council, founded in 1959 by the Ministry of Information and Social Services, along with some expatriates, as an independent agency. The objectives were articulated as follows:

> The council shall work towards the promotion, revival, development and encouragement of literary, visual and performing arts of Nigeria by: (a) assisting and encouraging deserving Nigerian artists and craftsmen; (b) initiating and participating in the

> revival, organization and conduct of cultural festivals, exhibitions, concerts and displays; (c) promoting and publicizing Nigerian arts in and outside Nigeria; (d) advising in the acquisition and preservation of art and cultural monuments; (e) fostering appreciation and pride in local history and monuments, the giving of lectures on local history and by education in the value of their artistic heritage; (f) encouraging the performance and exhibition of the forms of non-Nigerian cultural activities in Nigeria with a view to the development of indigenous arts by the impact of international culture. (Fasuyi, 1973, p. 39)

Activities in several other countries reflected similar objectives (see, for example, Chapter 4). Cultural centers were set up in cities to encourage traditional displays which might otherwise be constrained to rural areas, their ancestral bases. National dance troupes were established and sponsored.

Radio and television complemented these activities through live coverage of traditional activities, dramatization of traditional themes, and documentaries. In most African countries broadcast news is translated into several local languages (up to 20 in Nigeria alone). While the entertainment content of television programming in many countries was heavily European and American initially, the ratio increasingly swung in favor of local programming, including especially those that tap into folklore and social issues (Ibelema, 1980; Lasode, 1993; UNESCO, 1974). Television newscasters, whose visibility is a factor of cultural legitimation, are required to dress in traditional attire in several countries. Workplaces that forbade traditional outfits began to encourage, or at least accept, them soon after independence. Perhaps the ultimate expression of African commitment to African cultures was the Olympic-size Black and African Festival of Arts and Culture held in Senegal in 1965 and in Nigeria in 1977. These festivals brought together cultural performances in all of Africa and the African diaspora.

The universities, which used to be the bastions of Euro-modernity, have become centers for the study and promotion of African culture, with several now offering degrees in African languages, musicology, and arts. Whereas in the past "oral traditional poetry gave way to Shakespeare and English literature" (Fasuyi, 1973, p. 21), now all are studied distinctly or comparatively. Moreover, African writers, among them Chinue Achebe, Kamara Laye, Ngugi wa Thiong'o and Wole Soyinka have Africanized not just African literature, but also the colonial languages it is written in (Rhoads, 1993; Kalu, 1994).

The assertion of cultural integrity may be said to be one policy area in which African leaders and people are in accord. With the possible exception of Zaire, where Mobutu Sese Seko's "cultural authentication" policy (Bokonga, 1980) was rather intrusive and somewhat coercive, the post-colonial popularity of African cultural forms materialized either spontaneously or in reaction to encouraging policies. Today, African music, fashion, dances, hairstyles are in vogue, not just in primordial villages but in urban centers; not just among traditionalists, but also among the trendy youth. It should be noted that cultural identification with Africa by people of African descent—especially in the United States and the Caribbean—must have served as a catalyst for African youth's embrace of African culture as a contemporary lifestyle. External validation of African culture by people perceived to be in the centers of modernity can only enhance its appeal.

The emergence and popularity of contemporary African music is especially illustrative of the resilience of African cultures and the success of post-independent programmatic policies. In all regions of the continent, African melodies and lyrics were superimposed on a combination of Western and local instrumentation to produce music that is exuberantly rhythmic, that is, unmistakably African. This was the case in West Africa, where highlife music became a cultural mainstay epitomized by the works of Ghana's Ramblers Dance Band, and Nigeria's Victor Uwaifo, Rex Lawson and Osita Osadebe. From Central Africa, especially Congo, soukous music rose to global popularity by blending African melodies with the musical revelry of multiple guitars. Perhaps the ultimate blending of Western instrumentation and African culture can be found in the Yoruba's juju music, which seems to have conceded little of its traditional form despite the incorporation of Western instrumentation. The Western guitar seems merely to add an edge to the musical sonority, while the traditional talking drum rises unpredictably but incessantly to impose itself. In all three musical forms, African culture is preserved in tandem with innovation. (Africa's musical innovation is more fully discussed in Chapter 3.)

Even the doctrinal dichotomy between African traditions and Christian life has been significantly abandoned. As Kirby (1994, p. 67) states: "Today the majority of Africa's 230 million Christians . . . who were baptized in childhood and educated in Christian schools yearn for a return to their cultural roots and are fascinated by their traditional culture." Three general tendencies are evident in the adaptation to religion: Africanization of the mission churches, the fusing of African and missionary symbols and ethos, and modification of traditional spiritual practices. These changes, in fact, exemplify the general adaptation to the colonial experience and reflect profound cultural resilience.

Africanization takes the form of incorporation of African rituals, songs, and dances during services in mission churches or the establishment of new denominations or independent churches such as the Aladuras in Nigeria (Mudimbe, 1988). The first practice met with considerable resistance, especially by the Catholic hierarchy, but the attempt to discourage it was so futile that the Vatican was forced to issue a statement of acceptance, but with guidelines. In the fusing of church and traditional symbols, practitioners find commonalties between the two and basically equate one with the other. In one society, for instance, Christ is equated with the traditional nganga—an all-purpose spiritual healer, herbalist, and diviner (Schoffeleers, 1994, p. 75). Sachs (1979, p. 134) reports this instructive conversation between two Zimbabwean natives in the 1920s:

> When I go to church . . ., it is the Catholic church. I look at all the statues. I like to look at them and think they might be my midzimu [the spirit of one's dead relative].
>
> Yes, that is right. The same happens to me in my church. I pray, I think of my midzimu, and the faces change even if I don't close my eyes. The woman with the child turns into my mother, Nesta.

The modification of existing practices is arguably the most taxing as it involves a complicated process of evaluation, retention, and rejection. The cultural adaptation of an island village in Nigeria's Niger Delta illustrates this experience. Over the ages in Agbalama-Bonny, life, culture, and religion have revolved around the offerings and mystics of the rivers and the Atlantic ocean. Subsistence fishing has been the predominant traditional occupation, and cultural life is steeped in marine mythologies. Every month of August the masquerade club puts on a multiple-day display. The main attractions are masked dancers, or Owu, who are said to emerge from the depths of the ocean. For years, the Owu display has been one of the cultural landmarks of the village. Families return from cities and fishing ports to participate in it. The festival thus serves as a recreational outlet, a mid-season occupational recess, as it were, and an instrument of communal reaffirmation.

A less celebrated, if equally significant, display was the Kurualani, an annual mystic dance-festivity involving people (mostly women) who were said to be possessed by their marine spousal deities. They sang and danced animatedly, displaying agility not ordinarily associated with the individuals. Members of the Kurualani were also known as mystical healers. They were believed to have residual supernatural powers of diagnosis and cure passed on from their spousal deities. People—most of whom did not go to medical clinics—consulted Kurualani practitioners with ailments as divergent as infertility, malaria, and pulmonary disorders.

The Owu festivity and the Kurualani flourished for generations. But by the 1970s, they had begun to undergo significant and divergent transitions, reflecting socio-economic changes and epistemological adjustments. The Owu festivity remains a cherished part of social life, but as an increasing percentage of young men, who are its mainstay, became wage earners rather than self-employed fishermen, the festivity became less frequently held in August when there is no national holiday. It is performed more regularly, though in abbreviated forms, in December and/or early January to coincide with the Christmas/New Year holidays. The Kurualani, on the other hand, has been abandoned. Its members increasingly suffered credibility problems as more and more people questioned their claims of being possessed and of possessing supernatural powers. Their clients began to turn to medical cures. With its membership incrementally thinned through death and desertions and with few new "possessed" joining, the Kurualani festival soon folded.

It is important to emphasize that the basic ethos of the Kurualani remains alive and well. There are just not enough practitioners to make the annual group display feasible, or enough faithful to support a large group of practitioners. The transformation then is consistent with Obayemi's (1991, p. 22) argument that modernization is the expansion of options rather than the displacement of one with another.

The different fates of the Owu festival and the Kurualani, despite their common mythological roots, also illustrate Teheranian's (1977, p. 32) point that a "rational and innovative culture does not maximize reason or faith but optimizes both" through a process of self-examination and adjustment. The Kurualani was abandoned not because of political pressure or even overt persuasion, but through a process of epistemological adjustment, if not transformation. In due course, people had come to examine the record of the Kurualani, and, with the increasing availability of medical services, were able to compare it with the alternatives and to reject it. The Owu festivities, on the other hand, have remained as popular as ever among the youth and the aged alike. Like the Kurualani they are based on mythologies which few now believe, or perhaps ever believed. For instance, the Owu are said to emerge from the ocean depths under cover of darkness and return there after the festivities. But lack of belief in this myth has not reduced reverence for the Owu festival. Its rituals and codes of conduct are still strictly adhered to by everyone—for good reasons.

As a cultural landmark, the Owu festival has been a defining element of the people's lives. It is a link to the past that continues to serve the needs of the present: a reason to come home, and an occasion for communal bonding. Owu festivals serve the spiritual function of honoring the dead and providing a linkage to ancestors, whose current abodes are believed to be closely linked with Owuama (the land of divine masquerades). Thus, the utility of the festivals is not dependent on empirically verifiable realities.

The Owu festival also serves a maintenance function for the extended family. With economic necessities forcing members of the nuclear and extended families to separate by hundreds of miles, there is often the need to reassemble and reaffirm the kinship in ways possible only in the context of physical interaction. Owu festivals provide this context conveniently. Thus, the Owu tradition, like several similar traditions in Africa, serves to counter the dislocative effects of the modern economy.

However, the value of traditional practices goes beyond the rational, as indicated in Kluckhohn's definition of culture cited earlier. Accordingly, traditional practices of healing which may not survive empirical scrutiny have continued to thrive, even as an increasing number of people use the services of physicians and hospitals. J. O. Mume, a Nigerian entrepreneur in traditional medicine, claims that 80 percent of the people still see traditional healers (Mume, 1991, p. 314). Cursory observation of African societies would corroborate the thrust, if not the fact, of Mume's speculative figure. In recognition of this reality, several African governments have adopted policies that formally recognize herbal, mystic, and other traditional healers, who are now licensed in Nigeria, Ghana, and Kenya, among other African countries. African medical schools are also conducting formal research on herbal medicines with some promising results.

TRIUMPH TO WHAT END?

The title of this chapter suggests a finality in the outcome of Africa's cultural transformation. The reality, as has been noted throughout this analysis, is that the transformation is an on-going process. Indeed, as Owomoyela (1994) argues, there are still pressures, even from within the Africanist fold, to minimize the Africanness in African cultures or to portray it as a remnant of the past that in due course will dissolve in a universalist cultural order. Our view is that, African culture—in all its multiplicity—has survived and will endure and evolve. The pragmatic question though is, So what? Has the triumph of African culture made a difference?

In a general sense, the answer has been provided in the review of the notion of culture and transformation. In more specific ways, the question could be answered with reference to the relative stability and resilience of African societies, notwithstanding rampant civil wars, droughts, famines, and general economic stresses. For instance, the extended family system has remarkably resisted the pressures of modern economies, providing a buffer against some of the harsher realities. As Diop (1989, p. 177) writes, with ample poetic exaggeration, "[T]here are people who live in poverty, but no one feels alone and no one is in distress." In governance, Africa has been frequently decried for failing to develop democracy in the form of multiparty elections. It is only recently that the fallacy of equating electioneering with democracy has been widely acknowledged, at least with respect to Africa. It is not often noted that the more successful dictators or de facto presidents-for-life in Africa have been those who were adept at utilizing the informal processes of consultation and legitimation, including especially adherence to customary expectations. It has been noted that such consultations and representation sometimes guarantee a broader-based participation in the political process than does the voting process per se (Jackson and Rosberg, 1984; Wiredu, 1998). The point is not to minimize the significance of the electoral process in a modern society or to rationalize dictatorships. Rather, as Njoku E. Awa notes in Chapter 6, African traditions dictate democratic practices. Even military despots rarely last in office without reckoning with that cultural dictate.

It should be reiterated that Africa's cultural conservation has not precluded necessary changes (or has it?) In politics, for instance, reverence for elders and leaders continue even as Africans are increasingly voting out (or at least voting against) political icons such as Kenneth Kaunda in Zambia, Arap Moi in Kenya, and the late Felix Houphouet Boigny in Cote d'Ivoire. Communal bonds remain strong even as Africans are voting for reasons other than ethnicity, as in the election of Jerry Rawlings in Ghana, Nelson Mandela in South Africa, and Mashood K. O. Abiola and Olusegun Obasanjo in Nigeria. All four of these men won elections they should have lost if voting had followed patterns of ethnic loyalty. In Ghana, Rawlings won despite, opposition by the political leadership of the Ashantis, the dominant people of Ghana; in South Africa, Mandela faced an even fiercer challenge from the political leader of the dominant Zulus; and in Nigeria, unofficial published results of the subsequently nullified election shows that Abiola won in a contest against a member of the dominant Hausa-Fulani, garnering substantial votes in areas close to his opponent's base (How Abiola won, 1993). And Obasanjo was elected president despite losing in all states in Western Nigeria, his region of origin and ethnic domain.

Similar transformation is evident in social issues, including those of education, gender and health care. Formal education is now a passion even in societies where it used to be for the socially marginalized. In places, where girls were investment risks for educational purposes, an increasing number of women are sharing classrooms with their male counterparts (United Nations Development Programme, 1993). In nutritional practices, dysfunctional taboos such, as eating eggs, have been discarded and healthful habits formed (Igah, 1991; Dei, 1994).

It may be asked, Did Africa transform enough for its own good? If Weber (1958) is right in attributing the rise of capitalism (and hence Western economic advancement) to the "Protestant ethic," one might wonder whether, for instance, Africa's communalist values and life-celebratory ethic inhibited its economic advancement.

As Weber (1958, p. 17) defines it, the spirit of capitalism "is identical with the pursuit of profit, forever renewed profit, by means of continuous, rational, capitalistic enterprise."[5] This spirit is facilitated by the Calvinist work ethic, "the strict avoidance of all spontaneous enjoyment of life" (Weber, 1958, p. 53), and the "pessimistically inclined individualism" (p. 105), resulting in the accumulation of ever more capital as an end in life rather than the means of meeting material needs. This obsession is said to derive from the Calvinist belief that salvation is for a select few, and that perfection and success in one's calling is a sign that one is among the chosen (Weber, 1958, p. 133; Lessnoff, 1994, pp. 4–7). There's, however, the Marxian perspective that the Reformation movement as a whole was a religious compliment to the political and economic rise of the middle class, that is, a move to wrest authority from the Roman Catholic Church, in Frederick Engel's words, the "great international centre of feudalism" (Marshall, 1982, pp. 141–142). Indeed, Weber's thesis has generated considerable critique and dissension ever since it was originally published in 1905. If its validity is granted even to some degree, it is plausible to argue that the communalism and pleasure principles inherent in African culture may have hindered material development, as scholars such as Harrison (2000) and Harrison and Huntington (2000) have suggested.

But then, it is also possible that Africans could have turned Calvinist and, in the absence of the sufficient conditions, not derived the material gain. African countries' problems of structural adjustment (summarized in Chapter 8) points to this possibility. In that case, African culture might have become mired in the profit motive and lost the communal spirit and sense of pleasure that ease the harshness of material want. That, at a time when Western societies have increasingly come to reject obsession with pecuniary pursuits and to embrace a value system that sees material acquisition in its utility to self and society—or what may be called the African ethic. Kwame Nkrumah enunciated this ethic during his motion for Ghanaian independence, when he said:

> In our daily lives, we may lack those material comforts regarded as essential by the standards of the modern world, because so much of our wealth is still locked up in our land; but we have the gifts of laughter and joy, a love of music . . . . [W]hile we seek the material, cultural and economic advancement of our country, while we raise their standards of life, we shall not sacrifice their fundamental happiness. That has been the greatest tragedy of Western society since the industrial revolution. (Nkrumah, 1957, pp. 205–206)

To the extent that the preservation of African cultures has worked to prevent Nkrumah's fears, it has to be seen as one of post-colonial Africa's major achievements.

African cultures not only survived colonialism, but they are adapting and thriving. Their vibrancy today is testimony to their resilience and to the concerted efforts to reverse the effects of the colonial shock. Of course, there are new and continuing stresses from economic transformation and communicational hegemony. African cultures seem to be adapting to these, as well. With appropriate policies, African cultures should continue to evolve to meet the needs of the time and to facilitate Africa's material development without sacrificing Africans' humanistic ethos.

NOTES

1. In this regard, it is significant that "the African at the turn of the century regarded labor [that is, being a wage earner] as akin to slavery" (Gregory and Mandala, 1994, p. 222).

2. Despite this view which seemed to inform colonial policies, European explorers carted away African arts and artifacts to museums in London, Paris, and other former colonial capitals. African governments have repeatedly called for the return of the more significant artifacts, including the emblem of the Second World Black and African Festival of Arts and Culture (FESTAC '77), but to no avail (see Return the festival emblem, 1976).

3. These practices were in contravention of Article VI of the General Act of the Conference of Berlin (1885), which established the principles by which colonial powers were to relate to each other and conduct themselves in Africa. The paragraph on religious toleration states: "Freedom of conscience and religious toleration are expressly guaranteed to the natives, no less than to subjects and to foreigners" (excerpted in Fetter, 1979, p. 37).

4. As the Nigerian historian E. J. Alagoa notes, there is no documented evidence for this claim. Rather, it derives from folk etymology, which is widely accepted as authentic. However, "Some British visitors state that the name derived from brass neptunes the people used to manufacture salt" (E. J. Alagoa, personal communication, June 2002).

5. Weber takes pain to distinguish this definition of capitalism from the mere "impulse to acquisition . . . of the greatest possible amount of money" (p. 17), the difference being that the capitalist ethos stresses maximization of profit through rational calculations, rather than the maximization of income, say, through increased remuneration.

REFERENCES

Achebe, C. (1996). *Things fall apart.* Portsmouth, NH: Heinemann Educational.

Asante, M. K. (1990). Afrocentricity and culture. In M. K. Asante & K. W. Asante (Eds.), *African culture: the rhythms of unity* (pp. 1–12). Trenton, NJ: Africa World Press.

Awolowo, O. (1960). *Awo: The autobiography of Obafemi Awolowo.* Cambridge: Cambridge University Press.

Bokonga, B. E. (1980). *Communication policies in Zaire.* Paris: UNESCO.

Chinweizu (1975). *The West and the rest of us: White predators, black slavers and the African elite.* New York: Vintage Books.

Davidson, B. (1984). *The story of Africa.* London: Mitchell Beazley.

Dei, G. J. S. (1994). The women of a Ghanaian village: a study of social change. *African Studies Review, 37*(2), 121–145.

Diop, C. A. (1989). *The cultural unity of Africa.* London: Karnak House.

—— (1990). *Civilization or barbarism: An authentic anthropology.* New York: Lawrence Hill.

Eisenstadt, S. N. (1976). The changing vision of modernization and development. In W. Schramm & D. Lerner (Eds.), *Communication and change: the last ten years and the next* (pp. 31–44). Honolulu: University of Hawaii Press.

Fasuyi, T. A. (1973). *Cultural policy in Nigeria.* Paris: UNESCO.

Fetter, B. (Ed.). (1979). *Colonial rule in Africa.* Madison: University of Wisconsin Press.

Galtung, J. (1980). *The true worlds:* New York: Free Press.

Gergen, K. J. (1991). *The saturated self: Dilemmas of identity in contemporary life.* New York: Basic Books.

Gregory J. W. and Mandala, E. (1994). Dimensions of conflict: Emigrant labor from colonial Malawi and Zambia, 1900–1945. In D. D. Cordell & J. W. Gregory (Eds.), *African population and capitalism* (pp. 221–240). Madison: University of Wisconsin Press.

Harrison, L. E. (2000, Summer). Culture matters. *The National Interest*, pp. 55–65.

Harrison, L. E. & Huntington, W. P. (Eds.). (2000). *Culture matters: How values shape human progress.* New York: Basic Books.

How Abiola won. (1993, June 28). *TheNews*, p. 24

Ibelema, M. (1980). *Mass media in Nigeria: towards a development-support policy.* Master's Thesis, The Ohio State University.

Igah, S. A. (1991). Dietary habits, change and nutrition in Nigeria's economic development. In S. Bello & Y. Nasidi (Eds.), *Culture, economy and national development* (pp. 147–172). Lagos: National Council for Arts and Culture.

Isherwood, H. B. (1966). *Racial integration: The rising tide of colour.* London: 'Britons Publishing Co.

Jackson, R. H. & Rosberg, C. G. (1984). Popular legitimation in African multi-ethnic states. *The Journal of Modern African Studies*, *22*(2), pp. 177–198.

Kalu, A. C. (1994). Those left out in the rain: African literary theory and the reinvention of the African woman. *African Studies Review*, *37*(2), 77–95.

Kirby, J. P. (1994). Cultural change & religious conversion in West Africa. In T. D. Blakely, W. E. A. van Beek & D. L. Thomson (Eds.). *Religion in Africa* (pp. 57–71). Portsmouth, NH: Heinemann.

Kluckhohn, R. (Ed.). (1962). *Culture and behavior: Collected essays of Clyde Kluckhohn.* New York: The Free Press.

Lasode, O. (1993). *Television Broadcasting: The Nigerian Experience (1959–1992).* Ibadan, Nigeria: Caltop Publications.

Lerner, D. (1958). *The passing of a traditional society.* Chicago: Free Press.

Lessnoff, M. H. (1994). *The spirit of capitalism and the protestant ethic.* Hants, England: Edward Elgar.

MacMillan, W.M. (1949). *Emergent Africa.* Harmondsworth, England: Penguin Books.

Malinowski, B. (1944). *A scientific theory of culture and other essays.* Chapel Hill: University of North Carolina Press.

Marshall, G. (1982). *In search of the spirit of capitalism.* New York: Columbia University Press.

Mazrui, A. (1978). *Political values and the educated class in Africa.* Berkeley: University of California Press.

Mowlana, H. and Wilson, L. J. (1990). *The passing of modernity.* New York: Longman.

Mudimbe, V. Y. (1988). *The invention of Africa.* Bloomington: Indiana University Press.

Mume, J. O. (1991). Traditional medicine and development of modern pharmaceutical industry in Nigeria. In S. Bello and Y. Nasidi (Eds.), *Culture, economy and national development* (pp. 314–324). Lagos: National Council for Arts and Culture.

Ngugi wa Thiong'o. (1993). *Moving the centre: the struggle for cultural freedoms.* Portsmouth: Heinemenn

Nkrumah, K. (1957). *Ghana: the autobiography of Kwame Nkrumah.* London: Thomas Nelson

Obayemi, A. (1991). Culture in the Nigerian economy: some 1989 perspectives. In S. Bello and Y. Nasidi (Eds.), *Culture, economy and national development* (pp. 18–30). Lagos: National Council for Arts and Culture.

Owomoyela, O. (1994). With friends like these . . . a critique of pervasive anti-Africanisms in current African studies epistemology and methodology. *African Studies Review, 37*(3), 77–101.

Parker, E. B. and Mohammadi, A. (1977). National development support communication. In M. Teheranian, F. Hakimzadeh & M. L. Vidale (Eds.), *Communications policy for national development* (pp. 167–201). London: Routledge & Kegan Paul.

Return the festival emblem. (1976, October). *Africa*, p. 113.

Rhoads, D. A. (1993). Culture in Achebe's *Things Fall Apart*. *African Studies Review 36*(2), 61–72.

Rogers, E. M. and Shoemaker, F. F. (1971). *Communication of innovations*. New York: Free Press.

Sachs, W. (1979). Statement of a Zimbabwean non-believer. In B. Fetter (Ed.), *Colonial rule in Africa* (pp. 132–134). Madison: University of Wisconsin Press.

Schiller, H. I. (1976). *Communication and cultural domination*. White Plains, NY: International Arts and Sciences Press.

Schoffeleers, M. (1994). Christ in African folk theology: The Nganga paradigm. In T. D. Blake, W. E. A. van Beek and D. L. Thomson (Eds.), *Religion in Africa* (pp. 73–88). Portsmouth, N.H.: Heinemann.

Soyinka, W. (1990). The African world and the ethnocultural debate. In M. K. Asante & K. W. Asante (Eds.), *African culture: the rhythms of unity* (pp. 14–38). Trenton, NJ: Africa World Press.

Teheranian, M. (1977). Communications and national development. In M. Teheranian, F. Hakimzadeh and M. L. Vidale (Eds.), *Communications policy for national development* (pp. 1–50). London: Routledge & Kegan Paul.

Tehranian, M. (1990). Communication, peace, and development: a communitarian perspective. In F. Korzenny & S. Ting-Toomey (Eds.), *Communicating for peace, diplomacy and negotiation* (pp. 157–175). Newbury Park, CA: Sage.

United Nations Educational, Scientific, and Cultural Organization. (1974). *Mass media in an African context*. Paris: UNESCO.

United Nations Human Development Programme. (1993). *Human development Report 1993*. New York: Oxford University Press.

United Nations Human Development Programme. (1999). *Human development report 1999*. New York: Oxford University Press.

Weber, M. (1958). *The protestant ethic and the spirit of capitalism* (T. Parsons, Trans.). London: Unwin.

Wiredu, K. (1998). Democracy and consensus in African traditional politics: A plea for a non-party polity. In P. H. Coetzee & A. P. J. Roux (Eds.), *The African philosophy reader* (pp. 374–382). London: Routledge.

3

Artistic Creation in Post-Independence Africa

F. Abiola Irele

One of the most striking features of African life today is the heroic resilience of the population, the force of spirit which enables the ordinary people to confront the adversities of the post-colonial era. It is in the extraordinary context of political crisis and economic deprivation that prevails in so much of Africa today—of a comprehensive process of readjustment in the aftermath of the colonial experience—that cultural expression must be seen to assume significance: as a way of wresting meaning out of the confrontation with the conditions by which the existence of Africans is bound at this time. The very conditions seem, in their enormity, to have exerted a pressure on African creative faculties so that we have the paradox of a remarkable resurgence of creativity in a continent beset by a multitude of problems. When this paradox is considered in its full significance, we may well be justified in positing a cultural and moral wellspring from which such creativity flows. We are compelled to reflect further on the implications of this for cultural production in Africa when we consider that two major areas in which this creativity has been manifested are those in which a traditional inspiration has had to mobilize the resources of modern technology for its effective expression and dissemination: the cinema and music. This chapter discusses these two art forms in Africa, along with visual arts and literature.

CINEMA

The particular interest of the emergence of the cinema as a significant area of creative endeavor in contemporary Africa resides mainly in the fact that the medium represents a major innovation, involving the use of a complex technology, in the expressive culture of the continent. Moreover, African achievement in this area has been almost entirely a post-independence development. For although the earliest efforts to create an African cinema date back to the colonial period, the result cannot be considered to form part of an authentic current of African cinema. Such works as Nigeria's well-known film *Daybreak in Udi*, the product largely of British colonial initiative, or the ethnographic films of Jean Rouch, represent an early, extrinsic approach to cinema in Africa. It is only in the 1960s, with the entry into the field of gifted Africans, who were prepared to use film to narrate African lives and to project African realities, that African cinema, properly speaking—in the sense of imaginative engagement by Africans with the African experience, through the medium of film—can be said to have emerged. We might add that this emergence has been largely the result of a reaction both against the reductive trend of early colonial documentaries and the crude representation of Africa in western commercial cinema, typified by the Tarzan films and the infamous *King Kong* (Cameron, 1994; Dunn, 1996). As Frank Ukadike has remarked, "It is only after independence that the oppressive logistics of imperialism began to be investigated in cinema from an African perspective" (Ukadike, 1993, p. 46). The film that best illustrates this observation is perhaps Jean Marie Teno's *Afrique, je te plumerai*, with its graphic realism based on documentary footage related to the double harshness of German and French colonialism in the Cameroon.

The steady development of African cinema since the 1960s has witnessed a widening of its scope in such a way that the medium has come to encompass the full range of contemporary experience in Africa. This development, most notable in francophone Africa, was enabled chiefly by the work of the two great pioneers, Sembene Ousmane and Med Hondo. Ousmane's talent was first manifested in the simple lyricism of the short film *Borom Sarret*, based on one of his short stories. This film set the dominant pattern of his subsequent work, in which the emotional intensity of the narrative is felt as a direct function of its basic realist framework, as witnessed by such films as *Xala, Mandabi* (The Money Order), and *Thiaroye*. In these films, Ousmane's didactic intent and sense of social commitment is clearly foregrounded and is felt to predominate over any aesthetic ends that his handling of the cinematic medium may suggest. However, the 1976 film *Ceddo*, arguably Ousmane's masterpiece, departs somewhat from this pattern, in that its powerful critique of religious fanaticism is allowed to unfold in a wholly unforced way from a deployment of effects inherent in the cinematic medium itself. The narrative line is developed through the discreet use of flashbacks and graphic scene changes, while the camera is employed as a means of unveiling the inner life of the restricted circle of characters, each endowed with a powerful expressive life, most notably in the case of the Imam. The austere handling of its theme is thus sustained by a remarkable technical sophistication that recalls Kurosawa's *Rashomon*. The dramatic pace of his latest success, *Guelowar*, confirms the power of Sembene Ousmane's cinematic invention (Vieyra, 1972; Pfaff, 1984). As regards Med Hondo, the other giant of African cinema, we only need to mention his two outstanding works to indicate the significance of his contribution: the magnificent *Soleil-O* (1969) and the epic *Sarrounia* (1987), both of which reveal him as perhaps the most technically accomplished African filmmaker (Signaté, 1994; Cottenet-Hage, 1996).

The achievement of these two pioneers has been extended and consolidated by a group of gifted, younger filmmakers in francophone Africa whose work can be said to mark African cinema's coming of age. In *Yaaba* (1989), Idrissa Ouédraogo created a memorable fable of childhood imbued with a peculiar pathos related to the themes of age and social alienation. The element of fable is also present in Gaston Kabore's highly acclaimed first film, *Wend Kuuni* (1982), and in Djibril Diop Mambety's *Touki Bouki.* Mambety has followed up this work with the intensely realized work, *Hyenas* (1992). Billed as an African adaptation of the Swiss writer Dürenmatt's *Der Besuch der Alte Dame*, Mambety's film achieves a force of statement purely on its own narrative terms, in which the traditional element of fable, expressive of a communal morality, is given a visual treatment that exploits to the fullest the dramatic possibilities of the cinematic medium. In *La Petite Vendeuse de Soleil* (1999), the last film he directed before his untimely death, Mambety achieves a rare level of humanity that is given memorable and poignant expression in the final shot of the film.

Souleymane Cissé adheres to a remarkably effective simplicity of treatment in his film *Yeelen (*1980), a work destined to become a classic of African cinema as indeed of the seventh art itself. Here, the allegorical statement of the human condition, figured in the quest motif that structures the narrative, is developed through a mobilization of elemental images (fire and water, earth and sky), culminating in the powerful final image of a universal apocalypse that gives expression, in the appropriate terms of a nuclear explosion, to the abiding anxieties of human kind.

Apart from its intrinsic quality of figuration, *Yeelen* illustrates what is perhaps the most striking feature of these francophone African films, most of which have as their location the savanna of the West African Sahel: the way each one registers a vivid sense of place, generated by the visual dimension that defines the cinematic medium itself (Gardies, 1989). The narrative is no longer developed exclusively through the performance mode, as in the oral tradition, or through a text, as in the literate mode, but forms part of a total recreation of setting and action. In film after film, we witness the rituals of daily life in this area of Africa; in this way, we encounter the sights and sounds of the Sahel, its flora and fauna, captured to telling effect. The human context is seen as intimately linked to the vegetation, so that the flame tree, the shea butter tree, and especially the baobab emerge as distinctive features not merely of the physical environment but indeed, through their shaping of forms of life within it, as images of the social processes and cultural experience of the Sahel. Above all, the sparseness of the local landscape is given a compelling visual representation in these films in such a way as to provide direct pointers to their moral or symbolic content (MacRae, 1995).

Compared to the self-assurance of francophone productions, the cinema in anglophone Africa offers a less marked profile. The advantages enjoyed by francophone African filmmakers have not been available to the anglophones, except perhaps in South Africa, where both local conditions (in terms of financing and infrastructure), as well as international connections—as in the case notably of David Attenborough's involvement in the making of *Cry Freedom*—make it possible to envisage a national film industry as a viable proposition. These conditions do not exist in other parts of anglophone sub-Saharan Africa, nor can filmmakers count on support, in the form of subsidies by government and of financial investment and technical input by international organizations. However, some notable works have emerged, such as John Akomfrah's *Testament,* and Kwaw Ansah's *Heritage Africa*, made at great personal sacrifice. Mention should also be made of the Ethiopian-born American Haile Gerima whose film, *Sankofa,* has attracted attention and generated extensive discussion regarding its place not only in African cinema but also in black film generally. This is a work that needs to be placed in relation to black cinema in the diaspora, with reference in particular to the films of Spike Lee in the United States, to which we might add Julie Dash's *Daughters of the Dust* and Euzhan Palcy's *La Rue Cases-Nègres* (Sugarcane Alley). This raises the issue of the borders of African cinema, in terms that take it beyond questions of transnational transactions to consider its thematic orientations and ideological affiliations (Petty, 1999).

An interesting development, from the technical and socio-economic point of view, has been the emergence of the Yoruba cinema in Nigeria, as an outgrowth of the Yoruba Travelling Theatre. The transition of this theater movement from stage to screen was inaugurated by the veteran Hubert Ogunde, with the film *Aiye*. Ogunde was soon followed by another major figure of the Yoruba theatre, Moses Olaiya, with *Orun Moorun* (Ricard, 1983; Jeyifo, 1984). Because of the close links with television, the transition has now transformed itself into a substantial video industry in Nigeria, involving other ethnic groups, in particular Igbo. The same phenomenon can now be observed in Ghana, and in both countries we observe a divergence between the popular orientation of the videos and the economic imperatives of the video industry. Despite the predominantly middle-class clientele, who are able to afford both the hardware (VCR's and television sets), and the videocassettes, the orientation of the video productions has been towards a cultural fare, based on the Hollywood and Indian model, that is, distinctly popular in its appeal and references. This is in marked contrast to the situation in francophone Africa in which the cinema has been self-consciously oriented toward art, and is thus aimed at a more discriminating audience.

This brief review indicates the remarkable diversity of situations in which cinema is being produced by Africans, and of approaches to the art, even in its short history in the continent. This fact must incline us to caution in making generalizations about cinema in Africa. Nonetheless, we hazard two comments. The first concerns the thematic orientation of African cinema as it has evolved over the past 30 years or so. In this respect, Manthia Diawara has proposed a typology that recognizes three principal categories: social realist narratives, films related to the colonial confrontation, and those based on the return to the (African) source. As Diawara puts it, "Taken as a whole, it is clear that they [i.e., the films] reflect Africa in its quest for social and economic justice (social realist), identity (return to the source), and history (confrontation)" (Diawara,

1992, p. 142). These are valid categories as long as we do not overlook the considerable overlap between them.

Beyond this thematic classification, the important point that needs to be made relates to the novelty of the cinema as a medium and the consequences for African expression, for the imaginative impulse in Africa can be said to have found (with the cinema) a modern technological mode of expression. This development is all the more remarkable given the many technical and logistical problems that African filmmakers have had to contend with. There is indeed a sense in which the character of African cinema has been defined by the effort, despite these very odds to create a distinctive cinematic idiom. It is here that the theme identified by Diawara as a "return to the sources" assumes a particular interest, for this theme involves drawing upon tradition in terms not merely of content but also of the visual aesthetics by which such content is framed. The studied formality of much of francophone African cinema, often involving scenes shot in real time, certainly derives from an earlier mode of presentation established by the ethnographic film, especially in the work of Jean Rouch. Even so, there is a deliberate extension of this mode in films such as *Ceddo*, *Yeelen*, and *Wend Kuuni*, where an artistic rather than a documentary impulse is the guiding principle in the making of the film.

It is not too much to affirm therefore that the francophones have aimed at an art of the cinema that specifically draws its sanction from an aesthetics related specifically to an African sensibility. It is significant in this respect to note the way in which language is directly involved in the elaboration of an original cinematic aesthetic in francophone Africa. Indeed, one of the positive outcomes of this process has been the potential offered by the cinema for the revaluation of African languages, for their repositioning in the whole process of cultural production on the continent. The cinema offers a new form of narrative discourse based on a complex technicity. As Tomasseli has shown, the aesthetic and epistemological implications of this development for African cinema are far reaching; they call for nothing less than a revision of accepted western theories (Tomaselli, 1995). Without going into these issues, we might note that the technological basis of the cinema as an audio-visual medium introduces a singular dimension to the function of orality as a dominant modality of African expression, and further complicates the processes of communication involved in the complex interface between orality and literacy by which the experience of modernity has come to be marked on our continent. It is regrettable, however, that Tomaselli gives scant attention to the question of orality in his otherwise perceptive discussion of the issue of "cultural integrity" in African cinema. He touches only perfunctorily on the question of language as a cultural marker and dismisses orality as a valid criterion for the identification of African cinema, as suggested by Ukadike (1993, 1994).

MUSIC

The consequences of the application of modern technology to artistic expression in Africa in the post-independence period have been even more striking in the area of popular music, which has developed and flourished in the new urban centers that offer, in their physical settings, visible signs of our ambiguous modernity. Despite what may seem the negative implications of this association, popular music in Africa manifests an extraordinary vitality, the result of a long process of development that goes back to the inter-war years.

This development is well illustrated by the long evolution of this new form of music in West Africa. We get a sense of beginnings from the 1923 recording of Yaa Amposah and the 1929 recording of the group known as the West African Instrumental Quintet, in what was then the Gold Coast, later Ghana. Both of these recordings provide a vivid illustration of the progressive elaboration of the musical style that came to be known as highlife. The line of development is direct from these early efforts to the music of E. T. Mensah, who placed his individual stamp on this musical genre, and whose artistry helped to establish it as the dominant form of popular music in West Africa. His influence can be felt in the work of the various bands that sprang up in Ghana in the 1950s and 1960s, culminating in the polished and self assured style of the Ramblers in the 1970s (Collins, 1994).

The outstanding feature of this music is its syncretism, its easy blending of musical styles from various sources. The influence of church hymns is apparent in the melodic structure, while the predominance of brass instruments can be imputed to its early association with the brass bands that had arisen out of contact with colonial military music. Another major influence was the jazz music of the post-war period, especially that of Louis Armstrong, whose trumpet style is still dominant today in Ghana. Alongside highlife, and closely related to it, is the so-called guitar band music, played by such groups as EK's Band, Onyina, and Onyame, who evolved a performance style based on the folk genre *nwomkro*. Guitar band music has assumed a new dimension in the work of Nana Ampadu's African Brothers, whose combination of an appealing music with social and moral commentary related to the post-colonial situation in Ghana has greatly enhanced its popular appeal. The deliberate traditionalism of the group named Wulome in the 1970s represents an effort to broaden this appeal of popular music in Ghana through a return to indigenous sources. It is essential to remark on the role of the government owned radio station GBC 2 in the diffusion of this music.

The same influences and expressive processes were at work in Nigeria, where we can distinguish three main lines of development. Highlife was introduced in the early 1950s by Bobby Benson, and taken up by performers such as Zeal Onyia and Victor Olaiya, who gave the new musical idiom a distinctive Nigerian accent, and established a tradition that was sustained after independence by Victor Uwaifo, Celestine Ukwu, Rex Lawson, and Mbarga, whose phenomenal success, "Sweet Mother," defined the form for some time (Bender, 1991). The tradition continues to govern the music of Stephen Osita Osadebe. A second line of musical culture in Nigeria is represented by juju, a Yoruba genre that has come to be associated principally with the person and the style of Sunny Ade. Behind him, however, lies a long line of development, beginning in the 1940s with Akanbi Wright and Tunde King, and extending into the 1950s through the 1970s with Ayinde Bakare, I. K. Dairo, and Ebeneezer Obey (Waterman, 1990). The third line of development is embodied by Fela Anikulapo-Kuti with his "Afrobeat," the product of a process by which Fela gradually distilled jazz elements into a distinctly African form, while clarifying the relationship between the reference form (jazz) and its derivation (Afrobeat) on what a commentator has called "the stylistic, symbolic and functional levels" (Veal, 1995, p. 10). Fela's reworking of jazz to conform to an African temperament has been carried forward by his son Femi. It is a creative achievement that needs to be placed alongside the efforts of other African practitioners of jazz, especially in South Africa, among whom Hugh Masekela is certainly the best known figure.

Popular music has run a parallel course of development in other parts of Africa, especially in the Republic of Congo. Here, the predominant influence has been that of early recordings of Cuban *son* and *rhumba*, popularized through the length and breadth of Africa by the famous HMV GV label. This influence is evident in the early phase of Franco's career with his OK Jazz, and persists in the overall organization of his music to the end, as is demonstrated by his last masterpiece, "Mario." The same influence shapes the music of that other great figure of Congo music, Rochereau, whose transformation into Tabu Ley was announced with the traditionalism of his unforgettable number "Kaful Maye" and is in evidence in his more recent work, exemplified by the album entitled "Muzina." It is from these sources that Congo's soukous music emerged in the 1970s as a specific genre, with its own distinctive idiom, distinguished by its fascinating melodic and harmonic patterns and its rhythmic structure (Bemba, 1985). It is the social environment of this musical culture that is depicted in Mweze Ngangura's film, *La vie est belle.* As with highlife, we can speak of a tradition, represented by the music of such groups as Sakade (with their classic number, "Beya Moke"), Lipua Lipua, Sam, and, more lately, Lokito, Papa Wemba and Sam Mangwana. Perhaps the most significant figure today is Koffi Olomide. The attractive melodies, sophisticated harmonies, and original instrumentation of his early work announced a new approach to popular music in Congo. Along with these features, the tightly organized ensemble singing endowed his early music with an almost classic poise that still made room for a certain exuberance of expression. It must be said, however, that Olomide's later work has been very uneven, relying on tried formulas for increasingly meager effects.

The other notable musical form, Cameroonian *makossa,* introduced by Manu Dibango, has not had the same spread as Congo's soukous, though Pierre de Moussy's "Ndolo l'Amour" and "Ami O!" became widely known in their own time all over the continent. Working in another direction was Alpha Blondy of Ivory Coast, who counts as one of the foremost reggae musicians, giving a distinct African inflection to the form which in any case traces its origin to the African continent. The proximity of Congo to nearly all the nations on the eastern half of the continent, manifested in linguistic terms in the close relation between Lingala and Ki-Swahili, has ensured the penetration into these nations and the entire Southern African region of popular music inspired by or derived from

Congo's *soukous*. The forms and styles of popular music in these areas have nonetheless a recognizable character all to themselves, as is evident in the classic performances of the Western Jazz Band of Uganda, epitomized by their superb number "Vigelegele," or those of Thomas Mapfumo of Zimbabwe, whose traditional style, while recalling Tabu Ley, displays an originality of realization that is indisputable. The South African group known as Ladysmith Black Mambazo, whose a cappella style represents a development of the traditional form known as *isicathamiya*, is also notable for the haunting quality of their renderings and their original approach to popular music (Erlamann, 1996).

The dramatic expansion beyond Africa in recent years of the audience for African music has enabled it to occupy a central place in what has come to be called "world music," a development to which the forms of contemporary music from the West African Sahel have contributed significantly. This music goes back to the cultural ministry of Keita Fodeba, who formed the Ballets Guineens in the years immediately after the country's independence, and to the innovation of the Bembeya Band in blending the traditional griot repertoire with a new, modern style produced on western instruments. The musical idiom that resulted became a reference for urban musical entertainment in the region, and has been kept alive and renewed by the concerts and recordings of artists such as Salif Keita, Youssou Ndour, and especially, Toumani Diabate, and more recently, Ali Farka Touré. The music of the Sahel has led to a revaluation of its traditional background, for the use of western instruments has not diminished the importance of the two traditional instruments that determine its tonality and character, namely, the *balafong* and especially the *kora.* Diabate is without question the master of the kora, the greatest performer this instrument has known in living memory. His virtuosity is evident on his recordings, such as "Songhai" and its follow-up, "Songhai 2," both made with Western collaborators (Jose Soto in particular). His most accomplished work, however, is "Kaira" (released in 1987). Such is the brilliance and technical perfection of his performance on this CD that the recording company felt compelled to inform listeners that no special effects were used in the recording process, that what we hear on the record is entirely Diabate's musicianship.

As is evident from these examples, popular music in Africa is extremely varied. This music does not always display the complexity of traditional music, and it on occasion degenerates into mere boisterousness, or even a vulgar rowdiness, but it is almost always marked by vitality that is sustained by the instrumental mastery of its performers. It is no exaggeration to assert that African popular music is the most joyful music that is. This explains in part its appeal to non-Africans, which became evident with the rise to international stardom of Miriam Makeba, dubbed in the 1960s, "The Voice of Africa." Her success may well have aided the careers of such female figures as Mbila Bel and Angelique Kidjo; the emphatic manner and robust style of the latter seems to mark her out as the Amazon of African popular music. Kidjo reminds us of Tina Turner, whose style derives from the affirmative force of the blues. The rise to prominence in recent years of Cesaria Evora of Cape Verde confirms the trend in which female performers have enlivened the popular music scene in Africa.

No account of popular music in Africa is complete without a consideration of its impact on other aspects of musical expression, as part of the general movement of cultural production on the continent. Diabate's achievement has brought the *kora* to world attention. As Ravi Shankar showcased the Indian sitar to the world, Diabate's artistry has promoted an appreciation of the musical and expressive qualities of the kora, a remarkable African instrument. Kwabena Nketia has stated that, "as a melodic instrument, the kora stands high among African instruments." Nketia explains, "This is because of its wide compass, the contrasting sonorities of its twenty one strings that allow not only for variations in timbre but also for playing linear passages as well as simultaneous sounds" (Nketia, 1996, p. 84). The kora features prominently in "Keur Moussa," a collection of Wolof adaptations of Catholic liturgical music, and thus a Sahelian counterpart to the Bantu "Missa Luba." The kora is, appropriately enough, foregrounded in the opening and closing movements—"Libation" and "Epilogue"—of *African Portraits* by the African American composer Hannibal. This is a musical panorama of the African experience that begins with scenes from the ancestral homeland and takes us through the ordeal of slavery and into the struggle for survival in the diaspora. The various musical styles employed in the work serve as markers of the collective saga of the black race.

The mention of Hannibal's momentous work prompts a consideration of art music by African composers, a movement initiated by the pioneering effort of Fela Sowande, whose *Folk Symphony,* commissioned by the BBC to mark Nigeria's independence in 1960, represents the culmination of a distinguished musical career. His work runs parallel to that of Nketia in Ghana, and their conservative style, which aims to incorporate African folk material within the structure of the Western scale, in the manner of the so-called nationalist composers of central Europe during the nineteenth century, is in marked contrast to the modernist and experimental approach of a later generation, represented by the Nigerian, Akin Euba and the Ghanaian, Gyimah Labi (Irele, 1994; Mojola, 1995). However, Euba's major work, *Chaka*, an opera based on Senghor's dramatic poem of the same title, testifies to a new conception of art music in Africa, one that establishes a form of dialogue between the western symphonic medium and vocal styles, on one hand, and on the other, indigenous performance modes embracing poetic chant and instrumental music. This places in perspective attempts by outsiders to formulate an African idiom of art music, as in David Fenshawe's *African Sanctus*, and enlarges the fragmentary view presented by the Kronos Qaurtet in their best selling album *Pieces of Africa* or Philip Glass in his album *Screens.* Given these developments, it is heartening to observe the resurgence of African musicology in the work of Kwabena Nketia, extended in new critical and theoretical directions by indigenous scholars such as Akin Euba, Kofi Agawu, and James Makubuya. These scholars are providing us today with a new understanding of the formal structures of musical creation in various African cultures, and have inaugurated a trend that promises to lead to the renewal of musicology as a discipline, and thus ultimately to the revision of concepts of musical experience in general.

VISUAL ARTS

The achievements in cinema and music just reviewed are complemented and extended—one might even say, validated—by the creative upsurge in the visual arts that has marked cultural production in Africa in the post-independence period. Contemporary art has the special significance that in it the continuity between the indigenous tradition and the western-inspired conventions of artistic creation emerges most clearly. The obvious derivation from Yoruba sources of Ashiru's aluminum tableaux and Lamidi Fakeye's wood carvings, and of the prints and bead work of artists associated with the Oshogbo school, bears out this observation, which also applies to the work of that other Nigerian cluster of artists known as the Nsukka Group (Ottenberg, 1997). There is no question that the confident grounding in tradition lends force to the work of Muraino, who has emerged as one of the greatest artists of our time. The traditional inspiration is just as evident in the paintings, woodcuts, and sculptures of other Nigerian artists such as Grillo, Bruce Onobrakpeya, Demas Nwoko, and Uche Okeke.

Mention of the Oshogbo and Nsukka artists points to a noticeable feature of contemporary African art, the way it has developed within recognizable schools, often as a result of conscious promotion by individuals (in this regard, the connection of Ulli Beier with Oshogbo is well known) or official patronage, as is the case with the tapestries produced at the workshop at Thies. The splendor of these tapestries reflects not only the careful nurture of the workshop by Papa Ibra Tall, its first director, but also the personal interest shown by the Senegalese president, the poet Leopold Sedar Senghor, who sought to bequeath to his country through this initiative a national legacy of modern culture. It is easy to discern a common style in the famous stone sculptures of Zimbabwe, although the work of Euwitt Nyanhongo stands out clearly by reason of its boldness of conception and incisive technique. Similarly, among the work of a remarkable group of young artists that has emerged and appears to form a national school in the Republic of Benin, Julien Sinzogan's magnificent paintings are expressive, in their formal composition and vivid use of colors, of an artistic imagination that is decidedly uncommon.

These examples provide evidence of a powerful creative current in the visual arts in contemporary Africa. In the admirable album that the late Jean Kennedy devoted to this phenomenon, she rightly placed emphasis on the process of renewal and recreation to which it testifies, the productive interaction between the new and the old by which artistic creation is sustained on our continent at the present time (Kennedy, 1992). This has led to the elaboration of a new art that draws its power of conviction from its integration of traditional motifs and styles, while manifesting an intensity of expression that accords with a modern sensibility. As Ekpo Eyo remarked in his introduction to Jean Kennedy's collection, "African art has enjoyed a renaissance; an art of bold, new forms, full of diversity and vitality" (Kennedy, 1992, p. 11).

LITERATURE

To turn to literature is to begin to understand the importance of creative activity in post-independence Africa, for here the manifest discursive content of the literary text combines with the formal register to designate the symbolic and aesthetic as an area of contestation. As in the pre-independence period, African literature continues to bear directly in its thematic preoccupations on the conditions of existence of African peoples, on the various dilemmas that mark what Ali Mazrui has called "the African condition" (Mazrui, 1980). Given the dramatic and often agonizing circumstances of this condition, the compulsion to bear witness has often determined the dystopian character of much of this literature, and a corresponding mood of dissent. It is this critical impulse that has governed the direction of African writing in recent years that I have called elsewhere "the new realism," as opposed to the romanticism entailed by the literature of self affirmation that went with the nationalist phase of the African experience. What is remarkable about the new realism is the fact that it has fostered an unprecedented movement of stylistic innovation in the new literature, as evidenced by the novels of Ayi Kwei and Mia Couto, and especially by the francophone African drama, represented notably by the plays of Zinsou and Were Were Liking (Conteh-Morgan, 1994).

Given the negative mood that has pervaded so much of recent African writing, it is paradoxical that the difficult conditions of daily life have not impaired literary expression in Africa. Indeed, the establishment and rapid growth of local publishing houses have helped to promote the work of a younger generation of writers, a trend especially notable in Zimbabwe, where a veritable literary renaissance has been taking place since the 1980s, illustrated by the works of Chenjerai Hove, Charles Mungoshi, Tsitsi Dangaremba, and Paul Zinmuya, all initially published locally. The international success of the Zimbabwe book fair testifies to the prominent place assumed by that country in African publishing. A similar development can be observed in Nigeria with the poetry of Harry Garuba, Niyi Osundare, and Tanure Ojaide. As Paul Tiyambe Zeleza has noted, "It is remarkable that with limited resources, and against great odds, African publishers continue to publish work of high quality" (Zeleza, 1995).

It is thus fitting that African achievement in literature should be symbolized by striking success on the international plane. Wole Soyinka's Nobel Prize in 1986 and Ben Okri's Booker Prize in 1992 have both drawn attention to the continuing vitality of African literary expression. It is also important to stress that the mood of pessimism of the early post-independence period is gradually being dissipated in the consciousness of new possibilities in Africa, given trenchant expression in the works of Kofi Anyidoho and Niyi Osundare in poetry, and even a visionary elaboration in Ayi Kwei Armah's novel *Osiris Rising* and Kodjo Laing's *Achimota Wars* (Anyidoho, 1989; Irele, 1997). All this betokens a revival of hope in African literature, increasingly informed by the vision of a positive future for the continent. There is thus a real sense in which it can be said that utopia is back with Africans.

CONCLUSION

When the developments reviewed above are considered, we are led to the conclusion not only that the arts are alive and continue to flourish, but also that, despite the enormous difficulties we have had to face and endure, the creative spirit has not been diminished in its potency in Africa. It is in this light that we must reject the tendentiousness of the statement made a few years ago by Adewale Maja-Pearce: "The story of postcolonial Africa is the story of failure at every level" (Maja-Pearce, 1991, p. 5). This is the kind of sweeping remark to which the artistic achievements reviewed here provide a powerful refutation. This achievement contradicts the image that is being propagated of Africa as a continent mired in crisis and of its people as a humanity sunk in hopeless depression.

REFERENCES

Anyidoho, K. (1989). Beyond art: From dream to reality. *Asemka* [A literary journal of the University of Cape Coast] No. 6 (September), 19–31.

Bemba, S. (1985). *Cinquante ans de musique du Congo-Zaire*. Paris: Présence Africaine.

Bender, W. (1991). *Sweet mother: Modern African music*. Chicago: University of Chicago Press.

Cameron, K. M. (1994). *Africa on film*. New York: Continuum.

Collins, J. (1994). *Highlife time*. Accra: Anansesem Publications.

Conteh-Morgan, J. (1994). *Theatre and drama in francophone Africa*. Cambridge: Cambridge University Press.

Cottenet-Hage, M. (1996). Decolonizing images: *Soleil O* and the Cinemas of Med Hondo. In D. Sherzer (Ed.), *Cinema, colonialism and postcolonialism* (pp. 173–187). Austin: University of Texas Press.

Diawara, M. (1992). *African cinema: Politics and culture*. Bloomington: Indiana University Press.

Dunn, K. (1996). Lights . . . camera . . . images of Africa and Africans in Western popular films of the 1930s. *African Studies Review, 39* (1), 149–75.

Erlamann, V. (1996). *Nightsong: Performance, power and practice in South Africa*. Chicago: University of Chicago Press.

Gardies, A. (1989). *Cinéma d'Afrique noire francophone: L'Espace Mirroir*. Paris: L'Harmattan.

Irele, A. (1994, February). Is African music possible? *Transition* (New York), pp. 56–71.

—— (1997, April). African literature: past, present and future. *West Africa*, pp. 29–30.

Jeyifo, B. (1984). *The Yoruba travelling theatre of Nigeria*. Lagos: Nigeria Magazine.

Kennedy, J. (1992). *New currents, ancient rivers: Contemporary African artists in a generation of change*. Washington DC: Smithsonian Institution Press.

MacRae, S. (1995). Yeelen: A political fable of the Komo Blacksmith. *Research in African Literatures, 26* (3), 57–66.

Maja-Pearce, A. (1991). *Who's afraid of Wole Soyinka? Essays on censorship*. Oxford: Heinemann Educational Books.

Mazrui, A. (1980). *The African condition*. Cambridge: Cambridge University Press.

Mojola, B. (1995). *Nigerian art music*. Ibadan: IFRA.

Nketia, K. (1996). When times change. In M. Kopka and I. Brooks (Eds.), *Jali Kunda: Griots of West Africa and beyond* (p. 84). New York: Ellipsis Arts.

Ottenberg, S. (1997). *New traditions from Nigeria: Seven artists of the Nsukka Group.* Washington DC: Smithsonian Institution Press.

Petty, S. (1999). The archeology of origin: The transnational visions of Africa in a borderless cinema. *African Studies Review, 42* (4), 73–86.

Pfaff, F. (1984). *The cinema of Ousmane Sembene.* Westport, CT: Greenwood Press.

Ricard, A. (1983). Du théâtre au cinéma yoruba: Le cas nigérian. *CinamAction*, 26, 160–67.

Signaté, I. (1994). *Med Hondo: Un Cinéaste rebelle.* Paris: Présence Africaine.

Tomaselli, K. G. (1995). Theoretical perspectives on cinema in Africa: Culture, identity and diaspora. *Visual Anthropology*, 7, 297–329.

Ukadike, N. F. (1993). African films: A retrospective and a vision for the future. *Critical Arts: A Journal for Cultural Studies*, 7 (1 & 2), 43–60.

— (1994). *Black African cinema.* Berkeley: University of California Press.

Veal, M. (1995). Jazz music influences on the work of Fela Anikulapo Kuti. *Glendora Review*, *1*(1), 8–13.

Vieyra, P. (1972). *Sembène Ousmane, Cinéaste.* Paris: Présence Africaine.

Waterman, C. A. (1990). *Juju: A social history and ethnography of an African popular music.* Chicago: University of Chicago Press.

Zeleza, P. T. (1995). Acceptance speech, Noma Award, 1994. *Issue: A Journal of Opinion*, *23*(1), 7–8.

4

Cultural Politics in Post-Independence Senegal

Tracy D. Snipe

This chapter explores the promotion of culture and the arts in post-independence Senegal under the leadership of President Léopold Sédar Senghor and President Abdou Diouf. Senegal, known for its lively and influential intelligentsia, is perhaps the ideal francophone African country in which to examine the connection between politics and the arts since these two interconnected worlds have had an immediate and impressive impact on the character of the state's national and international identity during the era of independence. Equating culture with legitimacy, France pursued a policy of cultural assimilation in its colonies under the guise of "la mission civilisatrice." Assimilation emphasized French culture and the legitimacy of its rule. As a socio-cultural and political reaction to assimilation, the concept of negritude was developed among some educated elite from Senegal and other parts of French-speaking Africa and the Caribbean.

Like many other nations in sub-Saharan Africa, Senegal sought to unshackle the chains of cultural imperialism. Shortly after independence the state made several overtures toward the creation of a national culture. For example, the National Dance Company of Senegal (NDCS) was formed immediately after independence. Painters such as Papa Ibra Tall and Iba N'Diaye attained recognition abroad and played instrumental roles in the formation of art schools. Senegalese artists also distinguished themselves in filmmaking, the theater, and literature. Paulin Vieyra produced the first feature length film in sub-Saharan Africa, *Afrique Sur Seine*, in conjunction with several Senegalese directors including Mamadou Sarr, Jean Melo Kane, and Robert Christian (Ministère de la Cooperation et du Développement, 1991, pp. 315–316; Ndir et al., n.d., p. 37). Ousmane Sembene, a writer and director with whom Vieyra collaborated, has become recognized as one of the outstanding filmmakers in Africa. Douta Seck, the late Senegalese actor, was featured in Sembene's *Le Mandat* and Euzphan Palcy's *Sugar Cane Alley*. Talented female Senegalese writers such as Mariama Ba (*Une si longue lettre*), Aminata Sow Fall (*La grève des bàttu*), and Ken

Bugal (*Baobab fou*) have written critically and popularly acclaimed novels. Youssou N'Dour, one of the most famous artists in Senegal, enjoys an international audience because of the enormous appeal of Afro-pop.

During Senghor's presidency, the politics of negritude influenced policy. In essence, President Senghor wished to link Senegal to the intellectual and cultural West (M. Diouf, personal communication, November 22, 1991). Refuting European claims to the contrary, Senghor once described universal civilization as a product of value and cultural exchange. A significant objective of Senegalese cultural politics was to persuade the outside world to participate in this cultural exchange with Senegal and the African continent. Senghor wished to demonstrate that the black world had contributed to universal civilization despite the widespread perception that it lacked scientific or technological development—a fact which scholars such as Cheikh Anta Diop contest. Senghor was criticized for stating that "emotion is black . . . reason is Greek," which he never retracted (Mazrui, 1967, p. 9). Some of these arguments could be twisted to sustain the long-standing racist arguments of individuals such as the French philosopher Count de Gobineau.

Culture was the principal tool that Senghor used to come to terms with himself as a black African and French-trained intellectual and to fulfill his "mission" in life (Vaillant, 1990). Resurfacing in Paris in the 1930s at the height of the negritude movement, the notion of cultural assimilation was important in this struggle. Assimilation was a cultural and political question that Senghor continued to grapple with over the next few decades. As an intellectual, he may have stressed the need "to assimilate, not be assimilated," but as a statesman his policies were oriented toward assimilation (Senghor, 1964).

The tone of cultural policy in Senegal defies easy categorization, vacillating as it did between two poles: *enracinement* (rootedness) and *ouverture* (or opening). *Enracinement* and *ouverture* correspond to negritude and francophonie (M. Diouf, personal communication, November 22, 1991). Within the first decade of his presidency, Senghor's emphasis was clearly on enracinement, even though ouverture over the years grew more important for the Ministry of Culture. One of the central objectives of the bureaucracy at the end of colonialism was to reform existing cultural institutions and to create new structures that could better meet the needs of Senegal while preserving the cultural heritage of the nation (M'Bengue, 1973, p. 33). For instance, the state reorganized the Mali Arts Center to become the National Arts School, which served as a training ground for painters, sculptors, musicians and choreographers (M'Bengue, 1973, p. 35–36).

The Senegalese National Ballet epitomized the preservation of cultural heritage and promoted the principles of enracinement. Established within a year of Senegal's independence in 1961, the company selected the most talented dancers, drawn from different ethnic groups and regions, thus fostering a sense of national cultural unity. Guinea's Les Ballets Africains de Fodéba Keita was one of the first national dance companies in Africa, and President Senghor held it in high esteem (Senghor, 1964, p. 287). Maurice Sonar Senghor, nephew of the president, was one of the founders of the Senegalese national dance troupe, which has extensively toured Africa, Europe, the United States, the Soviet Union, and South America. The highly regarded Traditional Instrumental Ensemble was also affiliated with the Senegalese Ballet and the National Theatre (M'Bengue, 1973, p. 39). The state promoted these artists as Senegal's ambassadors abroad.

Aside from the fact that it fostered the principles of enracinement, the Manufacture Nationale de Tapisserie de Thiès deserves special consideration because it remains one of the few cultural institutions in Senegal that exist outside of Dakar. The tapestry manufacture was established by decree on February 19, 1966 (M'Bengue, 1973, p. 46). As a result of the attention that Senghor focused on the tapestries, some of them became gifts to heads of state. The manufacture produced cloth replicas of paintings by Senegalese painters such as Papa Ibra Tall, Ibou Diouf, Amadou Ba, Seydou Barry, Ousmane Faye, Papa Sidy Diop, and Mamadou Goudiaby (M'Bengue, 1973, p. 46). During the formative stages of cultural policy development, Senegal established various cultural institutions such as the Manufactures Senegalaises des Arts Decoratifs to further its national cultural objectives in preparation for the First World Festival of Negro Arts, an event that represented the apex of Senegalese cultural policy during the first decade of independence. This festival promoted the principles of enracinement and ouverture.

The Society of African Culture decided in Rome (in 1959) that a black African arts festival should be held in Dakar in 1966. Hosted by Senegal under the patronage of UNESCO, the First World Festival of Negro Arts was the crowning point of Senegalese cultural policy. It has had the greatest impact of any single cultural event in Senegal to date. The Musée Dynamique, an ocean-front museum located on the Corniche, was built for the first festival. Modeled on a Greek temple, the museum opened with "the largest museological exhibition of classical black African art to date" (Axt & Sy, 1989, p. 57). Art work from museums, cultural organizations, ethnic groups, and private collections from throughout the world were displayed. A Nigerian art exhibition was also held in the Dakar town hall. Quoting William Fagg, curator of the British Museum, Senghor proclaimed that "Nigeria is to Africa what Greece was and still is to the history of Europe. Nine out of ten of African carvings and works of art come from Nigeria" (Sane, 1977, pp. 7–8). Further, at an exhibition organized by the Senegalese painter Iba N'Diaye—"Tendances et Confrontations, les Arts Contemporains"—French writer and cultural critic André Malraux proclaimed, "You have got seven or eight painters of international distinction here who in no way fall short of the world's best" (Axt & Sy, 1989, p. 19). A seminal cultural event, the festival also clearly established Senegal as one of Africa's cultural leaders in the cinema. The festival awarded Ousmane Sembene a prize in filmmaking for La Noire de, the first full-length film by an African (Colloquium on Negro Art, Dakar, 1971, p. 221).

Senegal welcomed 2,500 artists and 25,000 guests, among them such luminaries as Duke Ellington, Katherine Dunham, and Langston Hughes (A Symbol of Continuity, 1977, p. 13). From an artistic and intellectual perspective, the festival was an event of Olympian proportions that revealed to the world the dynamism of African art—a fact immensely important psychologically (A. Cissé, personal communication, August 3, 1990). It would be impossible to sustain the argument that the festival served merely to assuage the bruised ego of the African intellectual and artist. Rather, it successfully showcased the very real achievements of blacks in the world of arts and letters, along with their unique contributions to jazz, blues, and spirituals, and many other fields of endeavor in the performing and fine arts, humanities, and sciences. The festival provided an extraordinary venue for black artists to demonstrate their achievements before a world audience. From this standpoint, its political and cultural importance remains unquestioned.

Senghor's reputation as a man of culture was personally enhanced, although Wole Soyinka called the Senghorian aspect of negritude into question during the festival. The future Nobel laureate from Nigeria proclaimed, "The tiger does not proclaim its tigritude. It leaps on its prey."[1] The remark created a controversy, revealing that many anglophone African intellectuals remain critical of negritude (Awoonor, 1975, p. 157). On the other hand, Abiola Irele, a Nigerian literary critic (and a contributor to this book), has commented favorably on the poetry of Senghor, calling him "the principal theoretician and apostle of negritude" (Irele, 1977, p. 1).

The national impact of the festival on the population is more difficult to access. A stadium was built in the suburbs to ensure that it would not be merely restricted to the commercial district. Yet most of the artistic events were located in Dakar and Gorée Island, so the large majority of Senegalese living outside the capital were unable to attend many events. Students attended some of the festivities (A. Top, personal communication, November 28, 1991), but the festival had the greatest impact on young painters and those involved in the theater and the production of tapestry (A. Bathily, personal communication, December 13, 1991). Cultural policy in Senegal has not benefited a national audience because the Ministry of Culture is centralized in Dakar. Major events such as the First World Festival of Negro Arts demonstrate that Senegal has the administrative capacity to promote and develop cultural programming, at least in the nation's capital. Radio has made music more accessible even in the rural areas. In terms of plastic arts and theater, regional cultural development remains virtually nonexistent outside of Dakar, with the exception of a few principal towns and cities such as Saint-Louis. Occasionally, the national theater tours the countryside but limits costly grand-scale productions to the capital.

Senghor continued to discuss enracinement until the waning years of his presidency, but several key cultural programs suggested the primary emphasis of Senegal's policy on culture had shifted to ouverture. The plastic arts had been promoted and funded so extensively that Senegal could now boast of its own School of Dakar in recognition of the virtuosity of its painters, though some critics noted Senegalese artists were essentially painting for a Western audience (Axt & Sy, 1989, p. 83). With the goal of promoting cultural exchange, in 1974 the state supported a traveling art exhibition of contemporary Senegalese art in the United States, Asia, and at the Grand Palais in Paris (Axt & Sy, 1989, p. 58). Ironically, artists eventually lost their training ground when the Musée Dynamique was converted to a dance school (Axt & Sy, 1989, p. 77).

The Centre Africain de Perfectionnement et de Recherche des Interprètes du Spectacle Mudra-Afrique also was based on the policy of ouverture. Popularly known as Mudra-Afrique from 1977–1984, the dance school was founded by Maurice Béjart, the director and choreographer for Bejart's Ballet of the 20th Century, and led by Germaine Ackogny (Axt & Sy, 1989, p. 58). President Senghor endorsed this school modeled after Mudra-Bruxelles (Acogny, 1980, p.

3). Mudra-Afrique, comprised of an international student body, was funded by Senegal and Belgium. The school emphasized the concepts of both enracinement and ouverture. Students received training in African, modern, and classical dance as well as drumming, and they performed in Dakar and the villages. Senghor did not clarify whether ouverture or enracinement should be stressed but was increasingly criticized for placing too much emphasis on the West. He defended his policies by stating that the two concepts shared a dialectical relationship, which according to Senegalese historian Mamadou Diouf (M. Diouf, personal communication, November 22, 1991) amounted to accommodating African culture and perspective to the West.

The maintenance of extensive cultural ties benefited both France and Senegal. Some critics maintain that under Senghor, the cultural policy of Senegal was unpopular and designed to reinforce French culture (L. Savane, personal communication, October 28, 1991). According to Carrie Dailey, "France probably was more comfortable influencing the cultural policy and the arts under Senghor because they were on the same wavelength" (C. Dailey, personal communication, November 14, 1991). French commercial, political, and cultural ties in Senegal were strengthened by organizations such as La Francophonie, the Ministry of Cooperation and Development, and Agence de Coopération Culturelle et Technique (ACCT), which provides scientific, technical, and cultural assistance to the French-speaking states of Africa (A. Bathily, personal communication, December 13, 1991). These agencies were important as external sources of funding for the arts and enlarged the sphere of French cultural influence in Senegal.

Language is one of the major ways that a culture is perpetuated. Although President Senghor was widely regarded as a Renaissance man, the six national languages (Wolof, Pular, Serer, Diola, Mandinka, and Sarakollé) were not taught in public schools or used in an official capacity during his presidency (A. Bathily, personal communication, December 13, 1991). Although approximately one-third of the population belongs to the Wolof ethnic group and Wolof is spoken by more than 70 percent of the population, French remains the official language of Senegal and proficiency in it still increases educational opportunity, access to the media, and career mobility in Senegal.

President Senghor was willing to finance a variety of cultural enterprises, which is ultimately the true test of his dedication to national cultural programming. The state has dispensed up to 25 percent of its budget annually for culture and education since independence (Axt & Sy, 1989, p. 20), but according to sources from the Ministry of Culture, never more than 1 percent of the budget was allocated for culture (A. Wele, personal communication, November 11, 1991). The amount of money specifically earmarked for culture was not very large, yet President Senghor spoke of culture almost constantly, keeping it firmly in the public consciousness. The sheer number of state cultural activities and the visibility of cultural programming were two of the key ingredients in Senghor's success as a policy maker. For example, La Loi Decorative reserved 1 percent of the total cost of public buildings exclusively for decorative purposes (based on Negro-African aesthetics) to enhance the architecture of the nation. (Ministère de la Culture, 1968, pp. 1–2.) In addition, at the ministerial level, President Senghor appointed individuals with impeccable credentials who shared his interest in the arts and education. Several cultural ministers attained further recognition in national affairs and in selected instances were recognized on the international level. Assan Seck became the minister of education, and later held the prestigious position of minister of state. Amadou Mahtar served two terms as director general of UNESCO. Alioune Sene was attached to the permanent mission of Senegal to the United Nations in Geneva.

President Senghor believed that culture was invented (M. Diouf, personal communication, November 22, 1991); he once referred to it as "the alpha and omega of development" (M'Bengue, 1973, p. 49). Culture was his way of putting Senegal on the map, and in the process he earned the title *mecière d'arts* or patron of the arts. The arts were never relegated to the sidelines during his presidency; they became an integral component of policy. With the possible exception of Julius Nyerere who translated several Shakespearean plays into Swahili, Kenneth Kaunda who advocated the concept of African humanism, and Kwame Nkrumah who stressed the concept of the African personality, Senghor supported modern cultural productions more consistently than any other African head of state. His retirement in 1981 meant a major cultural figure in Africa had departed from center stage.

Following President Senghor's voluntary resignation, Abdou Diouf, his former prime minister and personally chosen successor, became the second leader of post-independence Senegal. Prior to Senghor's retirement, in 1980 Senegal became one of the first countries to receive a structural adjustment loan from the World Bank. Structural adjustment emphasizes disengagement of the state and stresses the role of the private sector in the promotion of economic stability and growth. Senegal's economic decline continued unabated during the 1980s in spite of foreign aid that accounted for up to 20 percent of its gross national product, one of the highest amounts for the continent (Berg et al., 1990, p. i). The policy of structural adjustment had an adverse impact on cultural and social programs, which were given low priority under it (L. Savane, personal communication, October 28, 1991). Closer observation of cultural politics in Senegal during this period reveals that the failure of structural adjustment policy was rooted in politics. President Diouf and his cohorts became engaged in a protracted struggle to attain cultural recognition, at times manipulating traditional symbols for political advantage.

President Diouf immediately took steps to gain public acceptance of his new position and staked his claim to cultural legitimacy with Le Sursaut National and La Charte Culturelle National, a blueprint he established in 1985 for Senegalese cultural policy. Le Sursaut National[2] was initiated by technocrats or those members of the new generation of government leaders who had studied subjects such as computer science, law, and economics (M. Diouf, personal communication, November 22, 1991). Within the socialist party the debate between Sambou Cissé, a representative of Senghor's old guard, and Moustapha Ka, the major theoretician of Le Sursuat National, established a line of demarcation between individuals who remained faithful to Senghor's philosophy combining negritude and African socialism and President Diouf's new followers. President Diouf sought to shatter Senghor's legacy but could not devise his own cultural ideology, though his supporters borrowed heavily from the philosophy of Cheikh

Anta Diop, Senghor's principal intellectual rival (M. Diouf, personal communication, November 22, 1991).

By adopting Cheikh Anta Diop's cultural approach, President Diouf indirectly attacked the foundation on which negritude was built. Diop, a staunch cultural nationalist, favored Africanité, a cultural approach that emphasized national languages and cultures (Gray, 1989, p. 1). Pathé Diagne, among other intellectuals, played a key role in this regard, re-appropriating some of the central themes of Diop's philosophy to fit the context of a new discourse on the role of culture in national affairs and development. Shortly after Diop's death on February 7, 1986, a research institute with which he was affiliated, L'Institut Fondamental d'Afrique Noire (IFAN) was renamed in his honor. In 1987 the government renamed the Université de Dakar, Université Cheikh Anta Diop (Gray, 1989, p. 1). This honor was contrived to gain the sympathy of a public deeply moved by the loss of a renowned scholar and outspoken political figure.[3]

In contrast to President Senghor, who was diverted by ouverture, President Diouf emphasized enracinement, or the return to tradition. Unlike President Senghor, however, the technocrats did not approach the subject from an intellectual perspective, nor did they try to invent something new from the past. Instead, they merely rehashed long-standing ideas about oral traditions and beliefs (M. Diouf, personal communication, November 22, 1991). For instance, President Diouf came to office with his personal griot, El Hadj Mansour, who literally sang Diouf's praise before a national audience—for which he commanded a handsome fee. Mansour became one of the most celebrated commentators at the national radio and television station. The president, to some degree, based his legitimacy on Mansour who proclaimed him "Guardian of the Constitution" (M. Diouf, personal communication, November 22, 1991). President Diouf brought the position of the griot into further prominence in national affairs with the creation of the Comité National des Griots pour le Soutien à l'Action du Président Abdou Diouf (Diop & Diouf, 1990, p. 252).

Madior Boye, a woman judge, presided over the commission that created the plan for La Charte Culturelle Nationale or the National Cultural Charter. Notwithstanding pressure from artists and cultural elite, Mamadou Diouf remarked that implementation of the charter by the government was unlikely (M. Diouf, personal communication, November 22, 1991). He contended the charter was a calculated measure on the part of the government to undermine Senghor's cultural legacy. Abdoulaye Bathily, a historian at the Université Cheikh Anta Diop, observed that the cultural charter, never translated into practice, was merely an intellectual exercise for his colleagues who completed the necessary research and produced a written document (A. Bathily, personal communication, December 13, 1991). Bathily maintained that no one in Senegal could precisely describe what the cultural charter entailed or its importance to cultural policy.

Contrary to the repeated insistence of Minister Moustapha Kâ, no proposals from La Charte Culturelle Nationale were implemented. Minister Kâ cited *la journée culturelle* (cultural day) and *la semaine culturelle* (cultural week) as proof that the government had instituted some measures of reform (M. Kâ, personal communication, November 21, 1991). Mamadou Diouf who personally helped to draft the charter, disagreed with this assessment, adding that some of these programs had existed in the past (M. Diouf, personal communication, December 24, 1991). Neither Le Sursaut National nor the National Cultural Charter brought a new cultural policy or any significant break from the arts policy of the Senghorian era (A. Bathily, personal communication, December 13, 1991).

Senegal's reputation as one of Africa's cultural leaders was tarnished further in the late 1980s by a series of events that included the closing of the Musée Dynamique and the cancellation of the Pan-African Festival of the Arts and Culture (FESPAC), described as "Africa's largest cultural event since independence" (Brooke, 1988, p. 4). The government's decision to discontinue the museum (and use its building for a dance school, which was later converted to the Supreme Court) angered many artists. President Senghor had built Le Musée Dynamique for the First World Festival of Negro Arts in 1966. Its untimely demise was a blow to his cultural legacy at home and abroad. Festival organizer Pathé Diagne's brilliant but often controversial ideas were important in laying the groundwork for previous festivals. He was one of the principal organizers of both the Pan African Festival at Algiers (1969), which was critical of negritude, and the Lagos Festival (1977). Approaches based on the negritude and Pan-Africanist schools of thought were a primary source of conflict at these African arts festivals, resulting in numerous ideological and artistic disputes.

Diagne delayed FESPAC on three occasions before the government indefinitely postponed the festival on October 25, 1988, despite the fact it was scheduled to take place in Dakar the following June. An article in the *New York Times* attributed FESPAC's disintegration to the fact that Diagne's "organizational ability did not match his artistic vision" (Brooke, 1988, p. 4). According to Diagne, the *New York Times* supported the festival at the outset, but his opponents paid newspaper editors in Dakar to print stories that seriously undermined the organizers of the festival and dampened popular support for it. Diagne threatened to take the *New York Times* and *Jeune Afrique* to court for libel; neither source would print his rebuttal (P. Diagne, personal communication, July 11, 1990). The independent Senegalese press placed the blame squarely on Diagne and cited his mismanagement of festival funds—$2 million that was spent without any significant accomplishment (Brooke, 1988, p. 4).

Two years after the postponement of FESPAC, Senegal hosted the Biennal de Dakar, a state-sponsored cultural event in celebration of Senegalese and African literature. Numerous countries took part in this event, which included spectators and participants from Senegal and abroad. Lamine Sall, a poet and former president of the Association of Senegalese Writers, presided over the Biennal, which he judged to be a resounding success (A. L. Sall, personal communication, November 7, 1991). The Biennal received a less than enthusiastic response from those who regarded events of this nature as part of the government's attempt to attract more tourists to Senegal. The Biennal and gatherings like it are not elitist, even if a small segment of the population is directly involved in them; neither do they constitute cultural policy or a systematic promotion of culture by the government (C. Dailey, personal communication, November 14, 1991).

The Biennal de Dakar was part of the process of institutionalizing cultural events in Senegal, which reduced the influence of the arts and redefined the hegemony of culture in Senegalese society. Mamadou Diouf maintained that in the future Senegal would have fewer rituals of culture; every two years the state would organize a biennal, or significant cultural event, but it would last only several days. Such cultural events would no longer be significant because the promotion of culture would no longer be a government priority. The change was significant for a nation whose international reputation rested on cultural programming (M. Diouf, personal communication, November 22, 1991).

In 1990 the Diouf administration created presidential prizes for literature and the plastic arts to counter the widespread perception that artists were neglected under his leadership. Boubacar Boris Diop won the inaugural "grand prix littéraire du Président de la République pour les lettres" for his *Les tambours de la mémoire* (Biodan, 1990, p. 6). These awards represented Diouf's efforts to recognize the contributions of talented Senegalese artists but did not end criticism. The administration, albeit on a limited basis, continued to support filmmaking. In 1990 Rencontres Cinématographiques de Dakar (RECIDAK) was organized under the patronage of President Diouf to revitalize film production in Senegal. At the Pan-African Festival of Film and Television in 1991, Senegalese filmmakers were consistently surpassed by those from Mali and Burkina Faso. Short features directed by Djibril Diop Mambety (*Le Franc*) and Mansour Sora Wade (*Picc Mi*) have recently revived the cinema in Senegal.

State-sponsored cultural events such as *La Journee Culturelle de Sedhiou* and *La Semaine Culturelle* illustrate how the cultural policy of the Senegalese government had lost the dynamism it possessed under President Senghor. Those undertakings were reputed to promote folklore, but it was folklore with a political orientation. One observer lamented:

> We have gone from what I consider to be cultural activities to a lot of folklore. And the folklore is not even traditional folklore, the traditional cultural activities of the population. But the kind of folklore that you find involved in political activities. These events have become folkloric which has led to the reduction of genuine cultural activity in Senegal at present. (C. Dailey, personal communication, November 14, 1991)

Under French colonial administration, the Théâtre Ecole Normale William Ponty placed an emphasis strictly on folklore. Ironically, the Senegalese government has inadvertently reverted to financing such measures.

Without taking into account inflation, the budgetary allocation for culture and the arts under President Diouf was only slightly less than the amount appropriated by Senghor. Nevertheless, the public perception was that artists no longer received adequate funding. Individual artists may have suffered because they did not obtain as many subsidies, but the major differences in the change from Senghor to Diouf were absence of cultural leadership and Senegal's diminished visibility as a leader in the African arts internationally. President Senghor reserved a quarter of the nation's budget for education and culture, though a much smaller amount (1 percent) was specifically targeted for purely cultural activities. While low, this number is significant for a country in the developing world. Japan reserves O.068 percent of its budget for culture while France appropriates 0.86 percent for such programming (Governments, Censorship, and the Arts, 1989, sec. 2, p. 6).

Senegalese artists have always had to fight for their economic survival and independence. The arts policies of President Diouf and the implications of structural adjustment forced artists to look beyond the state for material support and to sustain themselves by becoming entrepreneurs (M. Diouf, personal communication, November 22, 1991). Several musical groups such as Touré Kunda developed outside state supported institutions. Despite the current reduction of governmental support for the arts, popular musicians such as Youssou N'Dour and Baba Maal thrived without state sustenance (L. Savane, personal communication, October 28, 1991). Albums and tape-recorded music are much less expensive than paintings, so artists have not fared as well.

Though President Diouf did not redefine *Africanité* or produce a major international cultural event, his administration provided support for several longstanding cultural institutions. He was one of the few African leaders to assist the Paris-based publishing house Présence Africaine in the midst of its financial crisis (C. Diop, personal communication, October 15, 1992). Diouf also continues to support the economically troubled African Cultural Institute located in Dakar. Notwithstanding such noteworthy contributions, Senegal's reputation in the artistic and intellectual domain cannot be solely attributed to its presidents, especially in consideration of the vast contributions made to discussions on politics and culture by Cheikh Anta Diop, Ousmane Sembene, Cheikh Amadou Kane, Aminata Sow Fall, and other prominent Senegalese artists and intellectuals who have immensely enriched the cultural life of the nation and contributed to the ongoing dialogue with politicians in a controversial but original manner.

NOTES

1. Wole Soyinka's comment illustrates the type of criticism the negritude movement has sustained over the years for romanticizing African culture.
2. *Sursaut* means start or big jump in French. For further elaboration on the significance of Le Sursaut National to Négritude, see, Diop and Diouf (1990, pp. 251–258).
3. According to Mamadou Diouf, scholars at the Université de Dakar first proposed that the government rename the university after Diop. The government rejected this initial demand, instead renaming IFAN, IFAN Cheikh Anta Diop. The university was not renamed after Diop until the following year.

REFERENCES

Acogny, G. (1980). *Danse africaine*. Dakar: Les Nouvelles Editions Africaines.

Awoonor, K. (1975). *Breast of the earth*. New York: Nok Publishers International.

Axt, F. A. & Sy, M. B. (Eds.). (1989). *Anthology of contemporary fine arts in Senegal*. Frankfurt am Main: Museum für Völkerkunde.

Berg, E. B. et al. (1990). *Adjustment postponed: Economic policy reform in Senegal in the 1980s*. Dakar: USAID.

Biodan, I. (1990, July 27). Lettres: Boubacar Boris Diop primé. *Le Politicien*, p. 6.

Brooke, J. B. (1988, November 13). Senegal postpones a black cultural festival. *New York Times*, p. 4.

Colloquium on Negro Art, Dakar. (Ed.). (1971). *Colloquium on Negro art*. Paris: Editions Présence Africaine.

Diop, M. C. D. & Diouf, M. (1990). *Le Sénégal sous Abdou Diouf: Etat et societé.* Paris: Editions Karthala.
Governments, Censorship, and the Arts. (1989, August 13). *New York Times*, sec. 2, pp. 1, 6.
Gray, C. (1989). *Conception of history: Cheikh Anta Diop and Théophile Obenga.* London: Karnak House.
Irele, A. (1977). *Selected poems of Léopold Sédar Senghor*. Cambridge: Cambridge University Press.
Mazrui, A. A. (1967). *Ancient Greece in African political thought*. Nairobi: East African Publishing Company.
M'Bengue, M. S. (1973). *Cultural policy in Senegal.* Paris: UNESCO.
Ministère de la Cooperation et du Développement. (1991). *Dictionnaire du cinéma africain. Paris: Karthala.*
Ministère de la Culture. (1968). Loi N^0 68-02 du 4 Janvier 1968. Relative à la decoration des bâtiments publics ou recevant du public.
Ndir, A. et al. (Eds.). (n.d.). Le cinéma sénégalais. *Senegal d'aujourd'hui magazine* [La Politique culturelle du Senegal], 37–39.
Sane, J. (1977). From Dakar to Lagos. *ICA Information*, Special FESTAC 77 (Dakar: African Cultural Institute), 7–8.
Senghor, L. S. (1964). *Liberté 1: Negritude et Humanisme.* Paris: Editions du Seuil.
A Symbol of Continuity. (1977). *ICA Information* Special FESTAC 77 (Dakar: African Cultural Institute), 12–14.
Vaillant, J. (1990). *Black, French and African: A Life of Léopold Sédar Senghor.* Cambridge, MA: Harvard University Press.

INTERVIEWS

Bathily, A. Interview by author, December 13, 1991.
Cissé, A. Interview by author, August 3, 1990.
Dailey, C. Interview by author, November 14, 1991.
Diagne, P. Interview by author, July 11, 1990.
Diop, C. Interview by author, October 15, 1992.
Diouf, M. Interview by author, November 22, 1991 and December 24, 1991.
Kâ, M. Interview by author, November 21, 1991.
Sall, A. L. Interview by author, November 7, 1991.
Savane, L. Interview by author, October 28, 1991.
Top, A. Interview by author, November 28, 1991.
Welé, A. Interview by author, November 11, 1991.

Part III

GOVERNANCE AND THE POLITICAL ORDER

5

Kinship and Civil Society in Post-Colonial Africa

Peter P. Ekeh

On the face of its common usage in African studies, the term *kinship* does not seem to have any heroic aura in its appearances. In any collection of terms intended to celebrate Africa's achievements in the post-colonial era, few Africanists would be inclined to include kinship among the glorious terminology of African history. It would probably remind many nationalist historians of a painful era of European imperialism that was intellectually dominated by colonial social anthropology and its inexorable quest for unearthing primitive kinship structures. In African studies, kinship has acquired an extravagant connotation of primordial connections based on blood relationships, real or imagined.

But it would be shortsighted to abandon kinship—a term that has come to capture so much of the ordinary in African history—to memories of historic humiliation. Unless the celebration of African achievements is to be limited to the high and mighty, which have been the fabric of African nationalist history, kinship must be reckoned to be well qualified for inclusion in the pool of constructs whose conduct has contributed significantly to the welfare of ordinary men and women in African history. Although it has come to assume a supra-individual cast, kinship's domain is the handiwork of ordinary men and women. It is a creation of ordinary people which, like community gods, another creation of theirs, now dominates their lives. Kinship and its institutions exist—and have survived in Africa—because they service the needs of ordinary people and protect them from various forms of victimization and the many dangers that abound in Africa's political circumstances.

Kinship in Africa owes its enduring value to its relevance for the lives of common persons. Contrary to its history elsewhere, in European civilization for instance, kinship in Africa has survived many historic changes because of its service to the common people. Indeed, it has grown bigger and more complex with the passage of time and various epochs. It is an institution that has withstood the onslaught of state tyranny—before colonialism, under colonialism, and in the post-colonial era. It is, therefore, the basic institution to which the ordinary person turns in times of danger. Indeed, it is fair to claim that historically, kinship has mediated the relationships between individuals and the state in Africa.

Herein lie the ties between kinship and civil society. Ultimately, civil society is a theory of individuals' freedom in the context of the behavior of the state. The structures of civil society are those whose operations in the context of the state maximize individuals' freedom. In African political history, kinship has operated to safeguard individuals' liberties from attacks from the state. It is therefore paramount that any assessment of civil society in Africa should be aligned with its interrelationships with kinship. Plotting a proper linkage between kinship and civil society is the task of this chapter in a book that draws attention to post-colonial African achievement. Kinship's achievements in protecting individuals from dangers are important and deserve some attention, even if they do not appear at first glance to be spectacular.

THE THEORY OF CIVIL SOCIETY

From its small beginnings in colonial social anthropology in the 1920s and 1930s, through its elaborate growth under later phases of colonialism and in the post-colonial era, the congeries of disciplines that have been grouped together as African studies didn't include the concept of civil society until the late 1980s. Nor was civil society a term of currency in the analysis of modern Western societies before this period. For most modern historians and political scientists, civil society was passe, serving only as a term used to characterize the historical evolution of Western institutions but of no great modern relevance. Since African studies has largely been dependent on Western theories and concepts for its intellectual livelihood, the notion of civil society was also absent from its conceptual repertoire for much of the existence of African studies.

Civil society blazed into renewed existence in the second half of the 1980s with the liberalization of politics in the Soviet bloc and the decay of the Cold War. Those who studied changing Soviet-bloc politics in Central and Eastern Europe patterned their analysis on Western European political sociology of the nineteenth century. Their intellectual quest was the maximization of individuals' freedom against the background of a history of the collectivization of politics in which individual liberty counted for little. For this mode of study the concept of civil society, culled from Western European history of the nineteenth century, was central to the presumed renaissance of individualism in the Soviet bloc in the late 1980s.

Civil society came into its own existence in African studies through the inspiration of these developments. Many post-colonial African states had flirted with Soviet-style command politics, with subsequent losses of basic freedoms that are usually associated with liberal democracy. Thus, some scholars of African politics saw a parallel development in the opening up of politics in both the Soviet Union and Africa in the 1980s, suggesting to them the prospects of a new order of civil society in Eastern Europe and Africa.

In the realm of the history of ideas, the quest for civil society has usually followed periods of absolutism in which individual liberties had been severely limited. Gellner (1991, p. 496) captured this point well:

> Twice since the Middle Ages has Europe been ideologically and politically bifurcated, with social order following the predilections of the rulers, on the *cuius regio eius religio* principle. The counter-reformation imposed a centralized, doctrinaire, intolerant, sometimes murderous, pervasive and self-righteous system on half of Europe, as Marxism was to do three centuries later. Absolutization of truth and centralization of power went hand in hand . . . . In each case, the countries finding themselves on the wrong side were condemned to somnolescence and torpor and retardation, and to an eventual effort at catching up.

In each case, the dissolution of such absolutism and centralization of power in Catholic Europe and Soviet Central and Eastern Europe has been followed by a new quest for individual freedom and its societal companion of civil society. But the structures that permit the emergence of individual freedoms vary enormously from regime to regime. While the common end of civil society is individual freedom, its structures of organization vary according to the historical circumstances of the regimes in question. The structures of civil society are not the same in post-Counter-Reformation Europe as in post-Soviet Central and Eastern Europe. We should not expect them to be the same in Africa as they are in Europe, even if they share a common historical necessity of seeking the benefits of civil society following a bout of absolutism.

It will be helpful for the understanding of the role of kinship in securing aspects of individuals' liberties in Africa to make a distinction between the functions of civil society and its structures.

FUNCTIONS OF CIVIL SOCIETY

Ultimately, civil society is a final cause construct, that is, its meaning emerges from the goal it seeks to accomplish. In this sense, the maximization of individuals' freedom is the ultimate quest of civil society. Black (1984, pp. 32, 43) puts it well: "The values of civil society comprise, first, personal security in the sense of freedom from the arbitrary passions of others, and freedom from domination in general . . . . We may identify the central ideal of civil society as personal independence and its central imperative as respect for persons."

The definition of civil society in its final cause terms allows us to see much more clearly the nature of the relationship between the state and individuals. Whenever the principles of civil society are operative, the welfare of individuals emerges as the ultimate goal of public policy. On the contrary, in illiberal regimes, individuals' efforts are harnessed to serve the promotion of some ultimate good or goal, that is not central to individuals' needs or that is even of no relevance to their welfare. Worse, in absolutist regimes, individuals and their welfare and freedom are often sacrificed for the purpose of advancing some ultimate goal of enhancing the needs of the state or of its functionaries. In absolutist and illiberal regimes, the goals of public policy are determined without any substantial input from ordinary persons.

A state with a functioning civil society operates a public domain in which the goals of public policy are negotiated and constantly redefined from the points of view of ordinary persons and their welfare. The public domain of illiberal polities excludes such debates, usually on the assumption that there exists some absolute truth from which individuals can derive satisfaction that will flow from their involuntary association with their society. Public domains wholly dominated by the state lack viable civil society because they tend to make individuals the instruments of public policy rather than its ends. Thus, the construction and control of the public domain are critical to the meaning of civil society. In order to achieve a stable society with a functioning civil society, the construction and control of the public domain must not be monopolized by the state; they should be shared between the state and other structures of society, which is why identifying the structural elements of civil society are important in assessing its health.

STRUCTURES OF CIVIL SOCIETY

Democracy and freedom were not always associated with civil society in Western European political thought. One of the earliest theorists of freedom in European history saw the intermediary structures of civil society as inimical to society's freedom. Starting a theme that informed much of the radical democracy of the French Revolution, Rousseau (1762, pp. 25–26) contended that:

> [W]hen cabals and partial associations are formed at the expense of the great association, the will of each such association . . . is *private* with regard to the State: it can then be said no longer that there are as many voters as men, but only as many as there are associations . . . . It is therefore of the utmost importance for obtaining the expression of the general will, that no partial society should be formed in the State, and that every citizen should speak his opinion entirely from himself.

The purging of the public domain of its autonomous "partial" associations, urged in Rousseau's doctrine, animated the initial operations of the French Revolution, including its draconian so-called *Loi Le Chaplier* of 1791 that baldly directed that "It shall not be permissible for citizens in certain occupations to meet together in defense of their pretended common interests. There must be no more guilds in the State, but only the individual interest of each citizen and the general interest" (cited in Bendix, Bendix, & Furniss, 1987).

Subsequent developments in Western political development led to the discovery that, far from encouraging and sustaining democracy and freedom, exclusive direct relationships between the individual and the state was inimical to personal freedom. Western political sociology has located the roots of democracy and individual freedoms in the existence of autonomous and free associations operating in the public domain alongside the state. It was to the robust existence of free and autonomous associations that de Tocqueville (1875) attributed the vitality of democracy and personal freedom in America (Kornhauser, 1959).

A wide variety of such associations has been identified in political sociology as being helpful to the existence of civil society and democracy in the West. They range from professional guilds and trade unions, political parties, and religious associations to organizations with narrower goals and purposes and shorter life spans. Their main attribute is that they operate in the public domain and share its resources, including its airwaves, without arbitrary control by the state.

ENEMIES OF INDIVIDUALS' FREEDOM IN AFRICA

The Western theory of civil society—emphasizing its central goal of providing ample room for the exercise of individuals' freedoms and the structures that enables them to operate in the public domain, without being subject to the arbitrary control of the state and its functionaries—enables us to explore the terrain and meaning of freedom in Africa. It also allows us to examine the turbulent relationship between ordinary Africans and African states, and kinship's ameliorating role in that relationship.

Beginning with the international slave trades from Africa by the Arabs (c.950–1850) and Western Europeans (1450–1850), there has been over several centuries a consistent attack on the liberties of ordinary Africans in an overwhelming attempt at exploiting their labor to service a developing international capitalism. African history owes an extraordinary amount of insight, which has not been fully acknowledged nor applied well enough, to Williams' (1944) *Slavery and Capitalism* for setting forth its brilliant thesis on this subject. Williams saw the trans-Atlantic slave trade as an integral aspect of the development of mercantile capitalism in England and Europe. The success of the slave trade and its companion mercantile capitalism allowed England to move to the second stage of capitalist development—industrial capitalism—that campaigned to terminate the overseas marketing of African labor, which it now needed for producing raw material inside tropical Africa for export to Europe in support of its engines of industrial production. Colonialism, according to Williams' thesis, was the product of industrial capitalism that needed organized African labor for producing the raw tropical products required for industrial production in England and Europe. Thus, both the slave trade and colonialism were organized victimization of ordinary Africans, which resulted in the loss of their freedoms to these international campaigns.

There was a domestic component in the campaign against ordinary Africans during the slave trade. With a few remarkable exceptions, African states were implicated in the persecution of ordinary Africans during the slave trade. Aside from a handful of states that declined to participate in it (Davidson, 1971, p. 65; Ekeh, 1990, pp. 676–77; Ryder, 1969, p. 198), most powerful pre-colonial states in Africa were converted into participation in the evil trade. In addition, a few states came into existence during (and from) the slave trade, with the primary purpose of servicing the needs of European traders for whom they acted as middlemen (Dike, 1956; Forde, 1956; Ikime, 1969; Jones, 1963). Just as success of the slave trade undermined the power of mercantile capitalism vis-a-vis industrial capitalism in England, inside Africa its consequences for indigenous states were devastating, weakening them to the point where they fell easy prey to their former European partners in the trade. Their misfortunes have been well noted in African nationalist history, although their perfidious role in the victimization of ordinary Africans has often been slighted in that history.

The loss of personal freedom under the slave trade was severe; however, individual freedom was not regained under the regimes that followed it, namely European colonial rule. The colonial state was a dictatorship. An astute observer of the colonial scene, the late Billy Dudley (1973, pp. 25–26) put it this way: "Colonial rule was, for all practical purposes, military rule . . . . Whatever legitimacy . . . the colonial authorities possessed derived . . . from the monopoly of the means of violence . . . . Colonial administrators were in fact 'praetors' and the system of rule was authoritarian and dictatorial." Personal freedoms were, indeed, meager under colonial rule.

Some colonial regimes in Africa represented more dangers to the welfare of ordinary individuals than what is often conveyed in such authoritarian and dictatorial systems of government. Some of them were murderous and most trampled on the human dignity of ordinary Africans. To cite the most egregious example of colonial plunder in Africa, Belgian King Leopold II's so-called Congo Free State (1885–1908) witnessed the butchering of millions of innocent Congolese that reduced the Congo's population drastically.[1]

Anti-Colonialism, Decolonization and Notions of Freedom and Democracy

Anti-colonialism denoted movements by Africans aimed at overthrowing colonial regimes, with an aspiration to replace them with indigenous African rulership. While varying in style and substance from region to region, anti-colonialism's operations spanned decades of the colonial experience in Africa. In Kenya, it was marked by the Mau Mau War (1948–53) waged by Kenyans against British settlers who had oppressed them and expropriated their lands. Bloodshed was also the signature of anti-colonialism in other British settler colonies in southern Africa as well as in Portuguese colonies. In British colonies in West Africa, anti-colonialism was primarily expressed in rhetorical denunciations of imperialism for which Nnamdi Azikiwe in Nigeria and Kwame Nkrumah in Ghana were famous. In the French colonies, anti-colonialism was mostly an intellectual enterprise, spawning such constructs as negritude which claimed special pride and virtue in being African, thus challenging French imperial claims of offering comprehensive and universal cultures.

Anti-colonialism was replaced in many European colonies in Africa by decolonization, a regime of cooperation between the colonizers and the anti-colonialists in attempts to transfer power to African rulers and to grant independence to the former colonies. Apart from affording opportunities to the colonial rulers and their African rivals to lay the groundwork for independence, decolonization also sought to reconcile differing visions, among rival African elite, of what national independence should portend and to offer arrangements for power sharing among various groups in the colonies.

In both anti-colonialism and decolonization, freedom and democracy were common terms used by the leaders of the independence movements who sought to wrench power from the colonizing Europeans. Nationalism in Africa in the 1930s and 1940s gained its intellectual current from a motivation to dethrone colonialism and promote in its place what nationalists regarded as its opposite, that is, democracy. The fiercest critic of European imperialism in Africa clearly saw the African situation in these terms. Nkrumah (1945, p. 25) complained that all colonial systems in Africa made it "impossible to vote the 'government' out of office," adding that they "also deliberately prevent[ed] and curb[ed] any aspirations towards independence on the part of the colonial peoples . . . . In fact, [they] definitely nullif[ied] the ideas of true democracy." Nkrumah (1945, p. 28) energetically affirmed that "it will not be incorrect to say that democracy, self-determination, independence and self-government are incompatible with the doctrines" of European colonial rule in Africa. In this earliest of his oft-repeated denunciations of colonialism, Nkrumah (1945, p. 43) offered a definition of a three-part end-product of a program of action that would lead to freedom and democracy in Africa as follows:

> (1) *Political Freedom*, i.e. complete and absolute independence from the control of any foreign government. (2) *Democratic Freedom*, i.e. freedom from political tyranny and the establishment of a democracy in which sovereignty is vested in the broad masses of the people. (3) *Social reconstruction*, i.e. freedom from poverty and economic exploitation and the improvement of the social and economic conditions of the people.

Defining colonialism as the antithesis of democracy was in fact a common practice among the nationalists agitating for independence from European imperialism in the 1940s. Another articulate nationalist of that era tied colonialism and democracy antithetically, even more sharply than Nkrumah. Ojike (1946, p. 271) condemned "British colonialism in Nigeria because . . . it contradicts the principles and practices of democracy. It makes one people to rule another under the guise of an advisory relationship."

Thus, in its earliest appearance in African affairs, democracy was an advocacy of freedom-fighting nationalism that sought to rid Africa of foreign domination.[2] Often paired with freedom, nationalism's democracy bore connotations that were radically different from the meanings of democracy in its native Western Europe from whence it was borrowed for use in Africa. This is clearly so on two separate counts. First, this early conception of democracy in Africa concerned itself with freedom from foreign rule. Its imprint was carried forward to the Fifth Pan-African Congress at which a statement prepared by Kwame Nkrumah was adopted:

> We believe in the rights of all peoples to govern themselves. We affirm the right of all colonial peoples to control their own destiny. All colonial peoples must be free from foreign imperialist control, whether political or economic. The peoples of the colonies must have the right to elect their own government, a government without restrictions from a foreign power. (Fifth Pan-African Congress 1945)

Stretching from its pioneering representations in Azikiwe (1937) to those of his younger and more militant colleagues, Nkrumah and Ojike, African nationalism's democracy was concerned with the freedom of peoples and collectivities from foreign domination and was silent on the problems of their freedom from domestic tyranny. Thus, Africans were ill-prepared by the advocacy of democracy in the nationalist era to tackle the problems of domestic tyrannies. In contrast, democracy developed historically in Western Europe in the context of the struggle for freedom from domestic repression. Although African nationalists had metaphorically copied the association between democracy and freedom from European political history, the contents and contexts of freedom were quite different in these two epochs.

There is a second difference between the native conception of democracy in the West and its adoption by African anti-colonial nationalism. In the modern West, the individual is the unit of operation of democracy. It is the expansion of the individual's legal, political, and social rights (Marshall, 1950) and the ordinary individual's enfranchisement that entrenched democracy in Western political culture. In contrast, African nationalism's advocacy of democracy was solely concerned with peoples' rights. It was largely uninterested in the rights of the unique individual. More likely, anti-colonial nationalists could not imagine that there was a valid difference between, on the one hand, the collective rights of a people to self-determination and, on the other, individuals' rights to possess their own niche of democratic expression concerning their rights of citizenship and freedom of speech.

These extraordinary conceptions of democracy during the nationalist era posed sharp problems for African political life precisely at the point of the triumph of nationalism. Basil Davidson, the Scotsman who was one of the few Britons who lent unwavering support to African nationalism, asked the frustrating question: "If nationalism has been and can be a liberating force, why then has it so often become the reverse?" (Davidson, 1992, p. 14). Nationalism's democracy ironically sowed the seeds of tyranny that came to wreck the prospects of individual-based democracy in post-colonial Africa. As matters turned out, a people's collective freedom could generate conditions that heighten lack of freedom for the individual. Nowhere is this irony more fulfilled than in the lives of the two most articulate spokesmen for nationalism's democracy in Africa. By sheer force of determination, Nkrumah broke down Britain's imperial maneuvers and won independence for Ghana in 1957. W. E. B. Du Bois, the African-American nationalist who pioneered pan-African nationalism across the Atlantic (Du Bois, 1945), followed the younger Nkrumah to Ghana as his principal adviser, apparently to live a fulfilled dream, in this land of freedom where he died in an independent African country in 1963.[3] Nkrumah soon came to be accused of being a tyrant who denied individuals the right of free speech. Nationalism's freedom from foreign rule might have been realized, but democracy's freedom also requires that the individual be free from domestic tyranny, a problem for which nationalist movements had not been prepared.

Africa's Post-Colonial States versus the Structures of Civil Society

Post-colonial states in Africa inherited the apparatus of colonial governments, together with their implements of violence, without the traditions of constitutional restraints that the existence of civil society compelled states to abide by in Western European nations. In the post-independence period, African states rapidly took over the total control of their nations' public domains. The major casualty of this new dictatorship of the post-colonial state was civil society. Hardly any conventional structures of civil society were left untouched by the sweeping sway of the post-colonial state. The dictatorship in post-colonial states ranged from personal rulership (for instance, in Guinea and Malawi) and military dictatorship (Nigeria, Ghana, etc.) to the so-called one-party democracies (Tanzania and Zambia).

The first structures of civil society to be brushed aside in all such instances of dictatorship was a competitive system of political parties. Other structures that perished, or at least were mortally weakened, in the destructive path carved by post-colonial dictatorial regimes included trade unions, student unions, and private newspapers. In virtually all of post-colonial African states, the public airwaves were monopolized by the state through state-run television and radio broadcasting. Similarly, the law courts and other devices of justice-making have been controlled by post-colonial African regimes.

Such a state of affairs paints a gloomy picture of personal freedom in Africa. Individual freedom would seem to have little chance of surviving in such suffocating circumstances. In all of post-colonial African states, individuals became mere instruments of public policy, hardly ever its end. This is evident from examining the least malignant of the post-colonial dictatorships in Africa. Tanzania's one-party state "democracy" is generally regarded by discerning Africans as being well-intentioned, even if it was mistaken. But even such a benign regime had a conception of the role of government that sacrifices the individual's freedom for the benefit of the state's development. As Nyerere tells it:

> In 1965 Tanzania adopted its own form of democracy—we rejected the Western model and said it was not appropriate for our circumstances despite the fact that all our constitutional development had until then been based on it. We looked at different democratic systems around the world, and studied the work of different thinkers . . . . Then we worked out a system of one-party Government which seemed to us to include the essential elements of democracy at the same time as it provided for unity and strength in Government, and took account of our poverty, our size, our traditions, and our aspirations. The resultant constitution is not perfect; but it suits us better than any system operating elsewhere, and *we believe that it safeguards the people's sovereignty at the same time as it enables the effective and strong Government so essential at this stage of our development.* (Nyerere, 1968, p. 19, emphasis added)

Note Nyerere's stress on people's collective sovereignty and the absence of any reference to individuals and their rights.

And yet there is an institution that has survived well in these painful circumstances, providing some measure of freedom and rudimentary elements of civil society for individuals in the face of agonizing attack on personal freedom in the civil domain. It is kinship which historically has sought to protect the individual in African history from the misbehaviors of the state.

KINSHIP AS PROTECTOR OF INDIVIDUALS' FREEDOM

African history is replete with the misbehaviors of states that have minimized the worth of ordinary people. The perfidious role of many indigenous African states in the international slave trade in reducing ordinary people in many communities to trading commodities; the depredations of colonial states against ordinary men and women; the outrageous behaviors of post-colonial states in monopolizing the control and uses of the public domain, marginalizing and endangering the lives of ordinary men and women in independent African nations: all of these have attacked ordinary Africans in ways and degrees that have broken the spirit of individuals in several other corners of human history.[4] Due to the operations of kinship and its institutions in the last several centuries of African history, the spirit of Africans have never been broken.

It is only by detailing and weighing the enormity of the perils that have haunted ordinary Africans in the last several centuries that the significance of the singular institution that has stood on their side can be assessed. African states—either through complicity with international economic forces or through sheer inefficiency and corruption—have been the main agency of individuals' misfortunes in African history of the last several centuries, at least since the sixteenth century.[5] On the other hand, kinship has played a leading role in providing some relief in the lives of ordinary people and a measure of dignity in their lives.

Kinship in Modern Africa

Conceptualizing kinship in terms that are comparable to the status of the state obviously raises important questions about just what kinship means in modern Africa. Defining kinship in modern Africa requires going beyond its anthropological meaning and recognizing its growing complexity and modernization in post-colonial Africa.

Social anthropologists imagined kinship as a complex of blood relationships, real or imagined, that bonded together self-conscious kinsfolk in a community (usually resident in a delimited locale) tracing its origins from a common ancestor. Colonial social anthropology's imagery of kinship in Africa was unfortunately static. There was the presumption that kinship was natural to Africa, varying little across the passages of time.[6] In fact, in the first few decades of the twentieth century, colonial social anthropology was studying kinship that had grown considerably in response to the ravages of the slave trade. Kinship was stronger in areas where the slave trade had been most intense or where it had claimed a large number of victims. For millions of beleaguered Africans, who had no protection from state organizations or who were persecuted by the state, kinship organizations became surrogate states during the centuries of the slave trade. It is sensible to push the hypothesis that kinship expanded enormously during the era of the slave trade in Africa, because it provided a vital source of survival for ordinary people (Ekeh, 1990, pp. 673–683).

Moreover, kinship has grown in scale and in its functions in response to the reorganization of African societies that European colonialism provoked during the last decades of the nineteenth century and during the first half of the twentieth. The security needs of individuals under colonialism were different from those that confronted them during the slave trade, but it was kinship organization, not the colonial state, that still assumed responsibility for ordinary people in colonial Africa. Where previously the individual's livelihood was confined to a small locality, in the colonial setting individuals had a larger space for their operations. Accordingly, kinship expanded along the routes of individuals' needs. Kinship, still understood in terms of blood relationships (now ever more fictionalized), has grown to include whole ethnic groups in modern Africa. Ordinary people in modern Africa have no problem arguing that those who speak the same languages and have common ethnicity have a right, or responsibility, to regard themselves as "brothers" and "sisters," offering one another protection and security if these are required. Their logic of reasoning broadens the meaning of kinship in modern Africa.

Kinship has also changed its meaning in a qualitative sense, especially in modern post-colonial Africa. Modern kinship in Africa is a series of informal corporations whose primary purpose is to provide the essential needs of its members and to protect its membership from dangers that stalk ordinary people in modern Africa. Because the state has failed to enter into these areas of their critical need, kinship has assumed responsibility for the birth and death rituals of ordinary individuals. The state has little to offer individuals as they approach death, nor is it responsible for the upkeep of those left behind by the deceased. These are kinship responsibilities. In other words, in modern Africa, kinship is responsible for essential welfare, for providing a safety net for ordinary people.

Kinship's Alternative Public Domains

One of the most remarkable social formations in modern Africa is the emergence of alternative public domains under the aegis of kinship systems. While the state dominates the civic public domain, crushing all those structures of civil society that challenge it within its arena, alternative public forums have emerged in which individuals express their cultural heritages and perform their political acts away from the overbearing presence of the state.

The relationship between the state and kinship in modern Africa needs careful delineation. In a critical sense, kinship gains its strength from the absence, weaknesses, or failures of the state. It is the failures and irresponsibility of the African state, historically and in modern times, that have emboldened kinship, compelling it to do for individuals what the state will not, or is not able to, do for them.

The origins of kinship's public domains, otherwise called primordial publics (Ekeh, 1975), must be traced to such contrasting relationships between kinship and the state. In many instances, the primordial public runs parallel institutions as the state. Ethnic associations and other branches of modern kinship in Africa have been known to run their own secondary schools and to institute their own financial systems to finance their public affairs, apart from the state's civic public affairs (Ekeh, 1994). But these are not standard public-spirited voluntary associations of Tocquevellian vintage that seek to boost the potency of an overarching common public realm. On the contrary, primordial publics fragment the overall public domain along narrow kinship lines.

Kinship's public domain contains structures that allow individuals to exercise their freedom within the political space of kinship systems. Individuals make rules that govern and bind them to a complex of duties and rights. Even more important, they provide a forum for moral behavior, a feature that is undermined by dictatorial regimes. Despised by the state and its functionaries, ordinary people enjoy a measure of dignity within the structures of the primordial public domain. Denied any room in the civic public domain managed by the state, ordinary men and women in Africa benefit from structures of civil society established within kinship's public domains, which range from women's and youth associations to credit associations. They are shielded from harassment from the state by their kinship labels.

A remarkable feature of modern Africa's politics is the autonomy of kinship institutions which African states have been compelled to recognize. This is so largely because kinship and its institutions have been defended by ordinary persons with a passion and loyalty that have not been extended to African states. Kinship's publics enjoy a degree of binding morality among its adherents that is lacking in the civic public domain.

CONCLUSIONS: LIMITS OF KINSHIP'S FREEDOM-MAKING POWERS

In a book consciously designed to celebrate success, such as this one on Africa's achievements in the post-colonial era, there is a built-in danger that the virtues of an institution like kinship may be overstated. From the point of view of the comparative analysis of kinship in Africa and elsewhere it may appear perverse that kinship is being cited in Africa as a freedom-giving institution. Its history elsewhere leads to very different conclusions. To cite a contrasting case, kinship withered away in Europe because it was no longer able to preserve security for the kinsfolk whose members had to look elsewhere for their protection and safety, leading to the rise of feudalism and, later, nation-states, that were able to provide greater security to individuals (Bloch, 1940, p. 142; Hintze, 1929, p. 27; Maine, 1888, pp. 155–156;). Moreover, individualism emerged in Europe precisely because ordinary people were freed from the shackles of kinship and village communities.[7]

Seen in this comparative light, kinship in Africa is indeed unique. First, far from withering away, it has grown along with the state, thus falsifying earlier expectations of modernization theory of the 1960s that kinship would die as the state developed. Second, kinship has cultivated important ties with individuals, providing them with essential welfare requirements, especially whenever the state proves unreliable. Third, kinship has in many instances sought to protect the individual from harassment from the state and has provided individuals important opportunities to exercise some limited freedom and human dignity. In other words, in the absence of services provided by kinship in these matters, the individual would be completely bereft of any modicum of freedom in many African states.

However, the level of freedom that the individual can attain from kinship is highly circumscribed; it has very little room for individualism that allows the expression of the individual's uniqueness. Custom is celebrated in the realm of kinship which in fact suppresses individualism in favor of community standards of behavior. Kinship allows far less than the amount of freedom available under liberal democracy that emphasizes the individual's uniqueness and his or her personal desires. Kinship only has room for its own members and lacks universal definitions of human conduct.

Moreover, kinship's form of freedom is qualitatively different from that attainable by individuals under a regime of civil society in a liberal democracy.

Kinship provides opportunities for *freedom from* persecution by the state and its agencies. What a mature civil society provides, on the other hand, is an opportunity for *freedom to* engage in activities that will enhance individuals' potentials for self-realization. For the attainment of self-centered freedom in Africa, as it has been in Europe, there is a need for both types of these freedoms to be realized. This means that the individuals will, ironically, need to attain *freedom from* the entanglements of kinship, which is only possible if the state changes its ways of relating to individuals by servicing their needs rather than alienating them from the civic public domain. It is only then that individuals will be able to exercise positive *freedom to* engage in activities that will enhance their potential. Meanwhile, kinship does well to provide some opportunity, however limited, for ordinary Africans to live in relative security.

NOTES

1. See Twain (1906); and Hannah Arendt's characterization of the racism that enabled imperialism to carry out offensive acts against Africans: "Two new devices for political organization and rule over foreign peoples were discovered during the first decades of imperialism. One was race as a principle of the body politic, and the other bureaucracy as a principle of foreign domination . . . . Race was the Boer's answer to the monstrosity of Africa . . . . [It] resulted in the most terrible massacres in recent history, the Boers' extermination of Hontetot tribes, the wild murdering by Carl Peters in German Southeast Africa, the decimation of the peaceful Congo population—from 20 to 40 million reduced to 8 million people" (Arendt, 1951, p. 185; see also Cookey, 1968).

2. In this regard it is noteworthy that Ghandi and Nehru's popularity among African nationalists was due to their success in winning political freedom for India, not because of India's success in the management of democratic practices in the post-independence era.

3. There are many in Ghana who believe that Du Bois' death robbed Nkrumah of an excellent source of wisdom and that Nkrumah's dictatorship would have been tampered by Du Bois' passion for individuals' liberties and his considerable influence on Nkrumah.

4. See Ekeh (1990, p. 693): "In general terms of world history, the centuries between the fifteenth and nineteenth bore traumatic experiences for the native cultures of Australasia, the Americas, and Africa. If there is judgment that in relative terms the African as an individual has withstood the turmoil occasioned by the slave trade and imperialism rather well, then this survival is owed more to the strength of kinship than to any state organization."

5. See Ekeh (forthcoming): "After Songhai's fall at the end of the sixteenth century [following its invasion by Morocco], the autonomy of African history began to unravel. After Songhai, the constant choice confronting African states was that between serving their people or joining in the international exploitation of ordinary Africans in a partnership between foreign interests and African states. That struggle came to define the character of African states and their relationship to their own citizens and neighborhoods. It was a dilemma whose resolution was weighted against the interests of ordinary individuals. A review of African history will show that after Songhai the bulk of the major indigenous states formed alliances with Arab and European interests to exploit ordinary Africans. In one sense, it was a matter of survival for these states. The trend for African states was to form these alliances with foreign adventurers who supplied them machinery for warfare, thus limiting their

reliance on their own citizens. African states that resisted this trend and insisted on their autonomy paid a heavy price for their assertion of independence. Benin City state, in what is modern Nigeria, provides a good illustration of this handicap. Insisting on the virtues of its illustrious history and the support of its citizens, the king of Benin and his government resisted British imperialism for a number of years at the close of the nineteenth century, at great cost. Benin 'now stood isolated as the last important traditional state surviving in southern Nigeria' (Ryder, 1969, p. 287) until its bloody sack by the British in 1897 (Bradbury, 1973, p. 91)."

6. See Ekeh (1990, p. 673): "Social anthropology apparently assumed kinship to be so natural to Africa that it saw no need to trace its sociological origins in other institutions or in previous historical epochs, nor to account for its persistence in the African historical landscape. Thus, social anthropologists studying African kinship systems (as the principal aspects of African social structures) immediately at the close of the slave trade did not relate them to the dynamics of that trade."

7. As Nisbet (1953, p. 107) puts it, "In the [European] medieval world there was relatively little concern with positive, discrete rights of *individuals*, largely because of the diffuseness of political power and the reality of innumerable group authorities. But when the consolidation of national political power brought with it a destruction of the social bodies within which individuals had immemorially lived and taken refuge, when, in sum, law became a more centralized and impersonal structure, with the individual as its unit, the concern for positive, constitutionally guaranteed rights of individuals became urgent."

REFERENCES

Arendt, H. (1951). *The origins of totalitarianism*. New York: Harcourt, Brace & World.

Azikiwe, N. (1937 [1968]). *Renascent Africa*. London: Frank Cass & Co. Ltd.

Bendix, R., Bendix, J. & Furniss, N. (1987). Reflections on modern western states and civil societies. In R. C. Braugard (Ed.). *Research in political sociology, Volume 3* (pp. 1–38.). Greenwich, CT: JAI Press.

Black, A. (1984). *Guilds and civil society in European political thought from the twelfth century to the present*. London: Methuen.

Bloch, M. (1940 [1961]). *Feudal society*. Chicago: University of Chicago Press.

Bradbury, R. E. (1973). *Benin studies*. P. Morton-Williams (Ed.). London: Oxford University Press.

Cookey, S. J. S. (1968). *Britain and the Congo question 1885–1913*. London: Longmans.

Davidson, B. (1971). Slaves or captives? Some notes on fantasy and facts. In N. Huggins, M. Kilson, and D. M. Fox (Eds.). *Key issues in the Afro-American experience* (pp. 54–93). New York: Harcourt Brace and Janovich.

—— (1992). *The black man's burden. Africa and the curse of the nation-state*. New York: Times Books.

de Tocqueville, A. (1875 [1966]). *Democracy in America*. J. P. Mayer & M. Lerner (Eds.). New York: Harper and Row.

Dike, K. O. (1956). *Trade and politics in the Niger Delta 1830–1885. An Introduction to the economic and political history of Nigeria*. Oxford: Clarendon Press.

Du Bois, W. E. B. (1963). The Pan-African Movement. In G. Padmore (Ed). *History of the Pan-African Congress* (pp. 13–26). London: Hammersmith Bookshop.

Dudley, B. J. (1973). *Instability and political Order. Politics and crisis in Nigeria*. Ibadan: Ibadan University Press.

Ekeh, P. P. (1975). Colonialism and the two publics in Africa: A theoretical statement. *Comparative Studies in Society and History, 17*(1), 91–112.

—— (1990). Social anthropology and two contrasting uses of tribalism in Africa. *Comparative Studies in Society and History, 30*(4), 660–700.

—— (1994). The public realm and public finance in Africa. In U. Himmelstrand, K. Kinyanjui & E. Mburugu (Eds.), *African perspectives on development: Controversies, dilemmas & openings* (pp. 234–258). London: James Curry.

—— (Forthcoming). Individuals' basic security needs and the limits of democracy in Africa. In B. Berman, D. Eyoh & W. Kymlicka (Eds.), *Ethnicity and democracy in Africa.* Oxford: James Currey Publishers.

Fifth Pan-African Congress. (1945). Declaration to the colonial peoples of the world. Reprinted in K. Nkrumah (1945 [1962]), *Towards colonial freedom. Africa in the struggle against world imperialism* (pp. 44–45). London: Panaf.

Forde, D. (Ed.) (1956). *Efik traders of Old Calabar. Containing the diary of Antera Duke, an Efik Slave-Trading Chief of the eighteenth century.* London: International African Institute.

Gellner, E. (1991). Civil society in historical context. *International Social Science Journal, 129*, 495–510.

Hintze, O. (1929 [1968]). The Nature of feudalism. In L. Frederick & F. L. Cheyette (Eds.). *Lordship and community in Medieval Europe. Selected readings* (pp. 22–31). New York: Holt, Rhinehart, & Winston.

Ikime, O. (1969). *Niger Delta rivalry: Itsekiri-Urhobo relations and the European presence. 1884–1936.* New York: Humanities Press.

Jones, G. I. (1963). *The Trading states of the oil rivers. A study of political development in Eastern Nigeria.* London: Oxford University Press.

Kornhauser, W. (1959). *The politics of mass society.* Glencoe, IL: Free Press.

Maine, S. H. (1888). *Lectures on the early history of institutions.* New York: Henry Holt & Company.

Marshall, T. H. (1950). *Citizenship and social class and other essays.* Cambridge: Cambridge University Press.

Nisbet, R. A. (1953). *The Quest for community.* New York: Oxford University Press.

Nkrumah, K. (1945 [1962]). *Towards colonial freedom. Africa in the struggle against world Imperialism.* London: Panaf.

Nyerere, J. K. (1968). *Freedom and socialism: Uhuru na ujamaa: A selection from writings and speeches, 1965–1967.* New York: Oxford University Press.

Ojike, M. (1946). *My Africa.* New York: The John Day Company.

Rousseau, J. (1762 [1957]). *The social contract.* New York: Hafner Publishing Company.

Ryder, A. F. C. (1969). *Benin and the Europeans, 1485–1897.* New York: Humanities Press.

Twain, M. (1906 [1961]). *King Leopold's soliloquy.* Berlin: Seven Seas.

Williams, E. (1944 [1964]). *Capitalism and slavery.* London: Andre Deutsch.

6

Participatory Decision Making in African Societies Before and After Colonization

Njoku E. Awa

An integral part of African societies since time immemorial, participatory decision making was sacrificed on the altar of colonialism. Because authoritarian colonial administrations wished to replace indigenous practices, customs, and commodities with Western methods, structures, and products, colonialism brought about an absolute change in African politics and society. Even African religious beliefs had to be converted to the "true" Western faith. In most cases, the surrogate culture that was imposed on Africans replaced an efficient and coherent model of society, a model that includes participatory decision making. In many instances, what Europeans brought to Africa were practices that were not only alien but socially inefficient, and African political culture suffered for this. Colonialism halted the growth of indigenous practices and, consequently, uprooted the very foundations of African political culture. Colonialism replaced traditional participation with an authoritarian decision-making structure. Yet, what remains of the traditional political culture is an impetus for Africa's postcolonial achievement in democratic reform.

A comparison of indigenous government machinery and Likert's (1961) models of decision making indicates that traditional decision-making structures were indeed participatory. Likert offers four systems of management: (1) the exploitative authoritative model, (2) the benevolent authoritative model, (3) the consultative model, and (4) the participative model. The exploitative authoritative model, rooted in classical organizational theory, pits workers and management against each other and is characterized by tight control. In the benevolent authoritative model, managers allow workers to voice their complaints and opinions, but workers actually contribute little because management does not heed their voice. In the consultative model, managers "actively seek feedback from workers, and use this feedback to direct organizational activities" (Kreps, 1990, p. 181). In the participative model, workers are encouraged to participate fully in organizational decision making and goal setting.

The first three models are pseudo-participatory; only the last is truly participatory. Likert contends that participative management or decision making is the most effective leadership style, similar to McGregor's (1960) position on the superiority of theory Y over theory X as a leadership style. We will see that, historically and currently, Africa offers numerous examples of an efficient and highly participative system of decision making, whose demise may well have contributed to development failures.

EXAMPLES OF PARTICIPATORY SYSTEMS IN AFRICA

The highly democratic and individualistic Ibo-speaking peoples of Nigeria reach decisions through participatory or group decision-making processes that closely approximate Likert's participative model. Ottenberg (1965) conducted extensive studies of the Afikpo, one of many Ibo-speaking peoples of Nigeria, and found them an autonomous and self-sufficient people. Afikpo council meetings allowed any member to voice an opinion on any matter brought up for discussion. No decision would be made until group consensus was reached; only when unanimous agreement could not be reached would the matter be brought to a religious authority (Ottenberg, 1965). The Afikpo government, moreover, employed a system of checks and balances to guard against the exploitation of power (Ottenberg, 1965). Even the traditional religious system was democratic. In addition, the Afikpo are organized by age grades. Each age grade consists of village-based groupings of people born within three years of each other and has a particular function. As one moves through the various grades, one's social duties change, for example from police work to legislative activities. Within each age group, members function as a democratic unit, using group decision-making and problem-solving methods to ensure the inclusion of diverse opinions and the probability of having solutions accepted.

The judicial system in Tiriki further illustrates group decision making in African traditional governance. The Tiriki, who are Bantu-speakers in the Nyanza Province of Kenya, have long institutionalized a form of democratic decision-making process (Sangree, 1965). A chief is assisted by a tribal council, selected through election by universal male suffrage. Such voting rights have developed since 1956. At the next level of governance—the subtribe (of which there are five among the Tiriki)—a headman who is appointed by the government confers with his sub-headmen before handing down decisions. Dialogue is crucial before either the headman or sub-headmen take any matter to the village chief. Such discussion opens channels for participatory decision making. In addition, the community still plays an important role as a judicial body. Grievances are presented to the elders of the society, and the community witnesses the trial and submits testimony as needed (Sangree, 1965).

The traditional lifestyle of the Baoulé people of Côte d'Ivoire further illustrates participatory structures for decision making in Africa. Father Vincent Guerry, a Benedictine monk who lived with the Baoulé, described them as a very communal people (Guerry, 1975). Through the group one is essentially given life, and life is given meaning by the village. Guerry describes what he calls one of the most important events for the Baoulé—the village meeting. During such meetings, various subjects may be discussed and everyone has a right to contribute. In this way the Baoulé live many hours together, as Guerry notes. Meetings continue at their own pace, with any village member making his or her voice heard. In this manner, a consensus is reached. The decision having been made by the group, the chief simply articulated what everyone present had already expressed. At no point had the chief interrupted the proceedings with his personal judgement. The people had spoken.

Guerry notes that the Baoulé are a proud people with a profound dislike for servitude. In everyday activity, they stress the importance of respecting personal independence. Everyone is entitled to an opinion. Thus, government is purely democratic, "as all larger decisions are made in public meetings where everybody, including women and youth, can have a say" (Guerry, 1975, p. 51). The chief and his group of elders represent authority. He is chosen through election by the people and may be removed by the pubic if he does not show respect for the Baoulé customs. Essentially, as a group of individual voices, the Baoulé village governs itself. Group cohesion is of paramount importance as the village gathers for the most important stages of life, including birth, sickness, and death. Perhaps there can be no clearer evidence of democratic decision making than the life of the Baoulé.

These are just a few examples of cases that have been systematically analyzed, but such participatory governance systems may be found in societies all across Africa. As the illustrations indicate, the democratic decision making and participatory development models which have been proposed for Africa are, in fact, part and parcel of the continent's cultural heritage. Africa's current democratization is rooted in that traditional political culture.

In his book *American Democracy,* Lipsitz (1986) notes that the concept of democracy has been defined in different ways, a spectrum from majoritarian democracy to egalitarian democracy. Lipsitz recognizes the basic tenets of democracy as majority rule, respect for individual rights, and a desire to reach a significant degree of equality among all people. He is also quick to point to the importance of popular participation.

It is quite ironic that while the West urges Africans to latch on to democratic ideals, Americans are decrying the democratic crisis in their own country because of nonparticipation, particularly at the polls. Meanwhile, in Africa, the spirit of participation is not moribund, despite its repression by a history of conquest and subsequent authoritarianism.

ORIGINS OF CENTRALIZATION AND AUTHORITARIANISM

The colonial period was a shock to African cultures, disrupting individual lives and entire societies and instigating rapid social change. Cross-cultural interaction is certainly enough to transform a worldview. However, the brutal nature of the European colonization of Africa has had extreme consequences in virtually every region of the continent and every sphere of life, including the political sphere. New political boundaries were drawn, lumping some people together and separating others, producing widespread political destabilization. Small European economic enclaves became large colonies as land was taken and divided, indigenous resources appropriated, labor redirected and forced, traditional authority denied, and humanity ignored. Throughout the process of colonization, Africans were unwilling but attentive students of European values and domination. Their learning would profoundly affect the shape that independent African states would take.

In *On The Theory of Social Change*, Hagen (1962) discusses the extreme psychological consequences that colonization has on the colonized. First comes force through outside intervention, which has an immediate effect on extant relationships. These relationships include economic and political ties, often defined through family and community connections. Traditional leaders were overshadowed by colonial administrators. They may have retained some token authority through the process of indirect rule, but in any case indigenous authorities "were left in nominal possession of their functions and prerogatives only if they submitted their exercise of power to the will of the conquerors" (Hagen, 1962, p. 412). Indigenous people were excluded from meaningful participation in the new political structures of which they had unwillingly become a part. Near-complete conquest and alienation, Hagen points out, can have only the most drastic consequences.

Second, there develops a kind of inferiority complex among the colonized. Among the manifestations is retreatism, what Robert K. Merton defined as an "abandonment of both the cultural and institutionalized norms of behavior" (quoted in Hagen, 1962, p. 195). Overwhelmed, individuals may appear simply to repress their goals and values and resign themselves to subjugation. This is usually interpreted as apathy.

In instances where submission meant near complete helplessness, Hagen states that one's reaction becomes more extreme and results in the personality type termed ritualist. People often took to identifying with their aggressors. Though they have no real chance of achieving the latter's status, they nonetheless continue to strive for this unattainable goal in a ritualistic manner. Out of this conflict were born the first African leaders of the post-independence period. "Sufficiently pursuing identification with the aggressors so that they acquired various European values, but just free enough so that they might speak out for independence, they gained a certain degree of charisma as leaders of the independence movement and so became leaders of the new nations" (Hagen, 1962, p. 420).

Caught between two value systems, however, many of these leaders were unable to completely identify with the African nationalist movement that aspired to regain Africa's cultural heritage. The result, in the case of presidents Mobutu Sese Seko in Zaire and Hastings Banda in Malawi, for instance, was a leadership of self-interest, which thrived under the guise of African nationalism and the fist of repression. Freedom of the press was denied in various countries in the name of national cohesion, to stamp out dissident voices in the fragile period of post-independence (Mytton, 1983, p. 36). In some cases, this led to a state of paranoia under which African leaders became intolerant of any criticism whatsoever. They ceased being accountable to their fellow citizens, and most people were denied meaningful participation in their own development.

The historian Basil Davidson (1994, p. 289) bemoans this situation by stating that the legacy of colonial rule left its unfortunate stamp on generations of African leadership:

> The forms that Europe has fostered here seem neither valid nor enduring. Rather do they now appear as an apparatus of helpless alienation in which creative potential is hard to find . . . . Yet go back in time and think about the past; and then there appears the outline of historical process and procedure which . . . had undoubtedly known how to control the use and the abuse of power.

In other words, as Davidson further expounds, this was a society where "effective participation could achieve a valid form of governance." Indeed, participatory democracy is increasingly becoming the model for Africa. To various degrees, this has been evident in Tanzania, Ghana, and Senegal. (See Chapter 11 in this book for a detailed discussion.) Similarly, Nigeria, Africa's most populous country, became a democracy again in 1999 after about 13 years of nearly uninterrupted military rule. All over the continent, there is a movement toward decentralization of economic and political power and a return to greater communal participation. And so it seems Africa is coming full circle politically.

THE POTENTIAL FOR CREATIVE REBELLION

Although it is not possible to start over in Africa, to wipe the slate clean of imperialism, Africans can reach back into various cultural repertoires for examples of efficient democratic decision making and to explore creative possibilities for the present and future. As indicated, this process has already begun. The previously mentioned example of participatory democracy through age groups still holds in contemporary Africa. In Abia State in Nigeria, especially in Abiriba, Ihe-Chiuwa, and Ohafia, age groups continue to function as localized examples of group decision making.

Another possible foundation for the democratic process in Africa is various traditional media, which play on participatory structures in conveying and sharing information. One example is the use of traditional theater in extension work. In Botswana, interactive theater has been used to present issues and ideas pertaining to development. Public performances always allow for audience feedback and thereby promote "participation and self-reliance in development by bringing people together to discuss their problems (reflected in the performance), agree on changes that needed to be made and take action" (Kidd and Byram, 1977, p. 1). By making use of media that have roots in participatory traditions, new messages may be incorporated while simultaneously allowing for public reaction and input, thus encouraging multiple voices in any decision-making process.

Africa has also witnessed the creative potential of combining traditional interpersonal and modern forms of communication to engage the public in debate and discussion. As an example, participatory radio forums have been implemented with some degree of success in countries such as Ghana and Nigeria (Bame, 1990, p. 140). In these forums information, generally on development related topics, is disseminated by radio to groups that then hold discussions about the content of the broadcasts. All those attending the forums are invited to share their feelings and ask questions about the information presented. This process turns passive listeners into active respondents. Once again, dialogue is vital to participation in decision making.

In Botswana, the government has undertaken an effort to build local institutions for agricultural development (Brown, 1990, p. 111). Since independence, the country has been responsible for the creation of a variety of local-level groups that receive considerable support and assistance. For Botswanan farmers, the government has sponsored both dam groups and drift fence groups. Though many local people have ignored these institutions, Brown believes that effective local agricultural groups "can provide farmers with a voice in policy decisions which affect them" (Brown, 1990, p. 111). Government willingness to provide these channels for participation is an encouraging sign for democratic growth.

After independence the southern African nation of Malawi was ruled ruthlessly for 30 years by Hastings Kamuzu Banda and the Malawi Congress Party (MCP). Instituting himself president for life, Banda in ways consistent with his authoritarian personality and his embrace of censorship defined government policies and national interests. This political situation began to change in 1993 as economic conditions worsened and Malawi looked for outside assistance. Almost overnight newly licensed publications began to appear. According to a correspondent for *Africa Report,* B. J. Kelso, "this proliferation of the press is symbolic of the democratic transformation under way in Malawi" (Kelso, 1993, p. 52).

As the international community focused attention on Malawi, a transformation took place. Independent and free journalism blossomed along with the multipartyism. Malawians were able to read and express alternative views in over a dozen new publications. Mutua (1994, p. 33) writes that the independent newspapers "energized the political process in Malawi," harkening back to "the dialogue of traditional participatory decision making."

In general, Africa's democratic rebirth, checkered though it may be, is undeniable. In a report entitled *Africa: Growth Renewed, Hope Rekindled,* the United States Aid for International Development (USAID) optimistically states that some 15 African countries were on the path to becoming "full-fledged multi-party democracies" (1992, p. 28), joining the fold of other African countries. Despite occasional movements in the opposite direction, such as the military coup in Cote d'Ivoire, the democratization process in Africa remains strong.

CONCLUSION

Africa does have something of its own on which to draw in the form of possibilities for popular participation in decision making. Channels must remain open for feedback and dialogue. As a participant at a London gathering of African women stated, "We must explode the silences of our people" ("Ourselves," 1988, p. 448). Where the African state has declared itself open to democratization, it is important to ensure that this does not mean democracy only for the few. Real participation has the potential to change entire countries, awaken dormant and distraught communities, and energize people to become decision makers for their collective future. This reality and potential are rooted in Africa's precolonial past, were repressed in colonial times, and are re-emergent once again.

REFERENCES

Bame, K. N. (1990). Some strategies for effective utilization of research information in Africa: Some Ghanaian and other African examples. In S. Arnold and A. Nitecki (Eds.), *Culture and development in Africa* (pp. 131–144). Trenton, NJ: New World Press.

Brown, C. (1990). Rural local institution for agricultural development in Botswana: No objection but no acceptance. In E. A. McDougall (Ed.), *Sustainable agriculture in Africa* (pp. 111–120). Trenton, NJ: New World Press.

Davidson, B. (1994). *The search for Africa.* New York: Random House.

Guerry, V. (1975). *Life with the Baoulé.* Washington DC: Three Continents Press.

Hagen, E. E. (1962). *On the theory of social change.* Homewood, IL: Dorsey Press.

Kelso, B. J. (1993). The press goes alternative. *Africa Report*, *38*(5), 52–53.

Kidd, R. & Byram, M. (1977). A Botswana case study: Popular theatre and development. *Convergence*, 10(2), 20–30.

Kreps, G. L. (1990). *Organizational communication.* New York: Longman.

Likert, T. (1961). *New patterns of management.* New York: McGraw-Hill.

Lipsitz, L. (1986). *American democracy.* New York: St. Martin's Press.

Mazrui, A. A., (1986). *The Africans: A triple heritage.* Boston: Little, Brown.

McGregor, D. (1960). *The human side of enterprise.* New York: Mcgraw-Hill.

Mutua, M. (1994). *Confronting the past: Accountability for human rights violations in Malawi.* Washington, DC: Robert F. Kennedy Memorial Center for Human Rights.

Mytton, G. (1983). *Mass communications in Africa.* London: Edward Arnold.

Ottenberg, P. (1965). The Afikpo Ibo of Eastern Nigeria. In J. L. Gibbs, Jr. (Ed.), *Peoples of Africa* (pp. 3–39). New York: Holt, Rinehart and Winston.

Ourselves as Women and as Africans. (1988, March 21). *West Africa*, p. 488.

Sangree, W. H. (1965). The Bantu Tiriki of Western Kenya. In J. L. Gibbs, Jr. (Ed.), *Peoples of Africa* (pp. 43–79). New York: Holt, Rinehart and Winston.

United States Aid for International Development. (1992). *Africa: Growth renewed, hope rekindled.* Washington, DC: USAID.

Part IV

ECONOMIC AND INFRASTRUCTURAL DEVELOPMENT

7

A New Look at Africa's Economic Growth and Prospects for the Twenty-First Century

H. Sylvain Boko

The economic literature is rich with studies attempting to explain global economic performance. The most often used methodology consists of an aggregative, reduced-form model, which attempts to explain the growth rate in real Gross Domestic Product as a function of variables reflecting (as required by the endogenous growth literature) the economic, social, political, and institutional environments of countries. With regard to African economic performance, the main conclusion of these studies, including Barro (1991, 1996, 1997), Easterly and Levine (1997), Collier and Gunning (1999), and Sachs and Warner (1997), is that economic growth rate in Africa is slower than that of other regions.

Explanations for this "significant Africa dummy" differ according to the authors. Easterly and Levine, who use the term "tragedy" to describe the slow growth performance of Africa, blame such factors as "low schooling, political instability, underdeveloped financial systems, distorted foreign exchange markets, and insufficient infrastructure," as well as ethnic fractionalization. In their survey article on African economic performance, Collier and Gunning (1999) conclude that there are three distinct causes of slow growth in Africa, namely, geography, macroeconomic policies, and microeconomic policies. Landes (1998, p. 499) adds the following to the chorus: "[A]ll the ills that have hurt Latin America and the Middle East are exponentially compounded in sub-Saharan Africa: bad government . . . poverty, hunger, disease, overpopulation—a plague of plagues."

Sender (1999, p. 90) observes, though, that the doomsday views of economic performance in Africa "is far too simple [in that] it fails to appreciate the complexity of processes of social and economic change in very poor economies." Consider the cross-country economic models used to analyze the African situation. Despite their comprehensive approach and their seemingly thorough investigation of the factors affecting economic growth, these cross-country and cross-regional studies fall short of their intended objective. First, by imposing the same model structure on all regions and all countries, the underlying assumption is that, with the exception of the rescaling of the regression constant through the use of regional dummies, there is a commonality in the factors explaining growth across all regions and across all countries. There is little justification for maintaining such an assumption.

Second, there is no standard on which to base the choice of the right-hand side variables to be included in the model. For instance, institutions in African countries continue to reflect their respective colonial heritage. The Beninese judicial system is highly reflective of the French one, and Kenyan law is based on British law. Further, there are important regional differences within Africa, which are left out in a cross-country approach. For example, while it is possible to analyze the effect of the Nigerian or the Ghanaian monetary policy on economic growth, the same analysis cannot be carried out in Senegal or Gabon, given that both of these countries belong to the African Financial Community (CFA), a monetary group of former French colonies in which member countries are bound by zone-wide policy on money and credit allocation. Further, in many countries, the institutional environment continues to evolve across time. In Benin for example, during the revolutionary period of 1972–1989, the banking system was completely monopolized by the government. Often credit was allocated, not on the basis of the merits of a proposed project, but on the basis of personal connections with government officials. The banking system has returned to private hands since 1990, and this, among other things, probably contributed to the economic recovery that has been experienced in the country since the early 1990s.

In all, the failure of aggregative cross-country studies to capture important country specific variables, and variability across time and across countries in factors that may affect growth limit the ability of these models to adequately explain growth performance in a given region. I concur with Collier and Gunning (1999) that both microeconomic research and case studies are necessary in the attempt to understand the growth performance in Africa. This chapter undertakes the latter suggestion.

The objective of the chapter is to showcase the efforts that have been undertaken in the area of economic policy and growth in African countries since the end of the colonial period. Because of limited scope and space, the analysis focuses on three African countries: Benin, Botswana, and Mauritius. The countries were selected to reflect differences in location, resource endowment, and political and economic structure. Benin, a former French colony, is located in West Africa, and is bordered to the south by the Atlantic Ocean. The country has few mineral resources, and its domestic production and exports continue to be dominated by agriculture. It adopted a democratic regime in 1990 after 18 years of military rule.

Botswana is a former British colony, located in Southern Africa. The country is landlocked, but is endowed with a variety of minerals, particularly diamonds, which constitute the bulk of its exports. Since its independence in 1966, the country has maintained a stable democratic system, with independent executive, legislative, and judicial bodies. Mauritius is an island republic in the Western Indian Ocean. It is a mostly mountainous country, with no natural resources, except its fertile soil. Of the three countries under study, only Mauritius has been successful at establishing an industrial base competitive enough to export its outputs to the rest of the world.

The chapter proceeds as follows. First, it provides an overview of economic progress in Africa since the mid 1990s in order to place the current analysis within the context of the recent positive general trends in economic performance throughout the continent. This is followed by case studies of Botswana, Mauritius, and Benin, respectively.

INDICATORS OF ECONOMIC RECOVERY IN AFRICA

Sub-Saharan Africa has often been described as the "economic basket case" of the world. After all, the often-cited numbers are daunting. As Table 7.1 shows, whereas the growth rate of GDP in low-income countries (as defined by the World Bank) increased from 4.8 percent annually over the period 1965–1980 to 7.5 percent in the 1980–1986 period, in sub-Saharan Africa the rate declined from 5.6 percent to 0.0 percent over the same two periods. With respect to growth in the components of GDP, agricultural annual output growth increased from 2.7 percent to 4.9 percent between the same two periods for low-income countries; in sub-Saharan Africa, the rate decreased from 1.6 percent to 1.2 percent. The contrast in the performance in industrial output was even more stark. While annual growth rate in industrial output increased from 7.5 percent to 10.6 percent in low-income economies, in sub-Saharan Africa the rate plummeted from 9.4 percent to –1.6 percent. Similar comparisons are applicable for the cases of the manufacturing and services sectors.

However, starting in the mid-1990s signs of economic recovery and economic progress can be observed throughout the continent, and these signs are general enough to warrant an optimistic view about the future of African economies. As reported in Bräutigam (1998), compared to the situation in 1990 when 15 sub-Saharan African economies experienced negative growth rates, in 1996, only 4 countries had negative growth rates. In the same year, the growth rate in Africa was actually higher than in Latin America. Over the three-year period of 1995–1998, the average growth rate of GDP was 4.3 percent in Africa. Comparing the period 1980–1990 to 1990–1997 (Table 7.2), the story is the opposite of that reported in Table 7.1. Indeed, the annual percentage growth rate of GDP in Africa increased over the two periods, from 1.7 to 2.1, but remained stagnant or even decreased in low-income countries (4.3 percent to 4.2 percent; they fell in middle-income countries (2.8 percent to 2.5 percent) and high-income countries (3.2 percent to 2.1 percent).

With respect to the structure of sub-Saharan Africa's economy, an important measure is the growth of value-added, which is a measure of the industrial sector's capacity for processing raw agricultural products into comestible or usable final products. In agriculture this measure, which grew at an annual average rate

Table 7.1
Average Annual Percentage Growth Rate in GDP and Its Components (1965–1986)

	GDP		*Agriculture*		*Industry*		*Manufacturing*		*Services*	
	1965-80	*80-86*	*65-80*	*80-86*	*65-80*	*80-86*	*65-80*	*80-86*	*65-80*	*80-86*
SS Africa	5.6	0.0	1.6	1.2	9.4	-1.6	8.5	0.3	7.5	0.1
LDCs	4.8	7.5	2.7	4.9	7.5	10.6	7.6	11.2	5.1	6.6

Source: World Bank, *World Development Report*, 1988.

of just 1.7 percent in 1980–1990, increased to 4.6 percent in 1990–1997. In South Asia the rate declined from 3.2 percent to 2.9 percent over those periods. Africa actually managed to increase its industrial output from 1.1 percent to 1.4 percent, although these numbers remain significantly below other comparable regions (World Bank, 1998/99).

Performance in other sectors also points to an economic turnaround. The annual growth rate in the exports of goods and services, which averaged about 2.2 percent in 1980–1990 more than doubled to a rate of 4.7 percent in the 1990–1997 period. In fact, over the period 1995–1998, exports from sub-Saharan Africa expanded at a rate of 10.6 percent a year. In contrast to the situation in the 1980s when continued deficiency in international reserves placed African countries in a precarious position with respect to their balance of payments, reserves have more than doubled since 1990. This economic turnaround is not limited to just a few countries. Indeed, 21 countries out of the 48 that constitute the sub-Saharan African region grew at rates of 5 percent or higher in 1996; and, in 1995–1997, 11 countries—among them the 3 countries under study—grew at rates above 7 percent. These statistics are remarkable when they

Table 7.2
Average Annual Percentage Growth Rate in GDP and Its Components (1980–1997)

	GDP		*Agriculture*		*Industry*		*Exports*		*Service*	
	1980-90	*90-97*	*80-90*	*90-97*	*80-90*	*90-97*	*80-90*	*90-97*	*80-90*	*90-97*
SS Africa	1.7	2.1	1.7	4.6	1.1	1.4	2.2	4.7	2.4	2.4
LDC's	4.3	4.2	2.6	3.5	5.3	5.1	..	..	5.0	5.6
Mid-Income	2.8	2.5	3.5	2.3	..	..	..	..	3.2	4.8
High-Income	3.2	2.1	2.2	0.8	3.2	0.7	5.1	6.7	3.3	1.9

Source: World Bank, *World Development Report,* 1998/1999.

are viewed in light of the seemingly desperate and regressive position the continent was described to be in just 10 years ago.

International private investors are taking notice of the economic rebound in Africa. After a long period during which the outflow of funds from Africa was greater than their inflow, foreign direct investment (FDI) to the region (while still only at a dismal 2 percent of the total FDI going to developing countries) is now growing; many equity funds now focus on Africa; and, venture capital is beginning to flow into the continent as well (Bräutigam, 1998). There is also a noticeable shift in economic policy toward Africa in Western capitals. The main example of this is the passage in 2000 of the African Growth and Opportunity Act by the U.S. Congress. This Act purports to open U.S. markets to African-made goods and encourage private investment into the region.

It is in the context of this recent positive economic performance throughout the continent that the case studies below are set. The case of Botswana follows.

BOTSWANA

The Country, the People, the Land, and Its Resources

Botswana is a landlocked southern African country, bounded by Namibia on its north and west sides; on the northeastern side by Zambia and Zimbabwe, and in the south and southeast by the Republic of South Africa. The country covers an area of 600,372 sq. km, an equivalent of 231,805 sq. mi. Its population is about 1,325,920. The Tswana constitute the vast majority of the population. Other major ethnic groups include the San, the Kalanga, and the Herero. The adult literacy rate is above 74.4 percent (United Nations Development Program, *Human Development Report*, 1999).

Rainfall varies from 25 inches in the north to 9 inches in the south, but as evidenced by the negative effects of the country's worst drought in the century in 1991, the economy remains vulnerable to drought. The country is abundant in mineral resources, particularly diamonds, which account for 70 percent of its exports earnings. Other minerals include copper, coal, cobalt, manganese, salt, asbestos, and nickel.

The Government

Botswana is a constitutional democracy. Its constitution was promulgated in 1965, a year prior to its independence from Britain. The National Assembly is the supreme legislative body consisting of 40 members elected to five-year terms by universal adult suffrage plus four specially elected members, the speaker and the attorney general, who does not have voting power. The National Assembly elects the president, who is assisted by a vice-president and a cabinet of about 10 ministers in his executive duties. The judicial system includes magistrates' courts, the High Court and the Court of Appeal. The political system in Botswana has a unique feature in that traditional rulers form an integral part of modern governance. The House of Chiefs—comprising 15 members—must be consulted by the government on all matters related to the chieftancies and constitutional changes. Vaughan (forthcoming) observes that "powerful Tswana chiefs as rulers of local communities not only exert considerable influence in the electoral system, but were also critical to the implementation of important state policies since independence in 1968."

Economic Growth and Policy Since Independence

The economic performance of this small country of just over a million people is nothing short of phenomenal, as shown in Figure 7.1. Upon independence from Britain in 1966, the country's real GDP per capita (calculated in 1990 prices) was $688, making it one of the poorest nations in the world. By 1997, the country's real GDP per capita had climbed to $6,057, placing it firmly among the middle-income countries. The growth rate in real GDP has remained strong, consistently averaging around 10 percent annually throughout the 1970s and the 1980s. This strong economic performance was fueled mainly by the country's exports sector (Figure 7.2), which is dominated by the sale of diamonds, which accounts for 30 percent of GDP and 70 percent of exports revenues. In the 1970s the average growth rate in exports revenues was about 30 percent; in the 1980s about 12 percent.

The dependence on diamond exports may put into question the sustainability of the country's economic performance, but the country will likely adjust appropriately. Indeed, in addition to the stable and democratic political system, which has enabled economic agents to operate in a credible and predictable environment, another remarkable characteristic of the nation is the restrained policy stance pursued by the authorities over time. In contrast to other mineral-rich and/or oil abundant countries on the continent, Botswana's authorities have avoided major white elephant projects and corruption. Instead they have harnessed the revenues from the sale of diamonds to improve the welfare of the whole country. They have directed these resources into: (1) productive public

Figure 7.1
Real Per Capita GDP in Botswana Since 1960

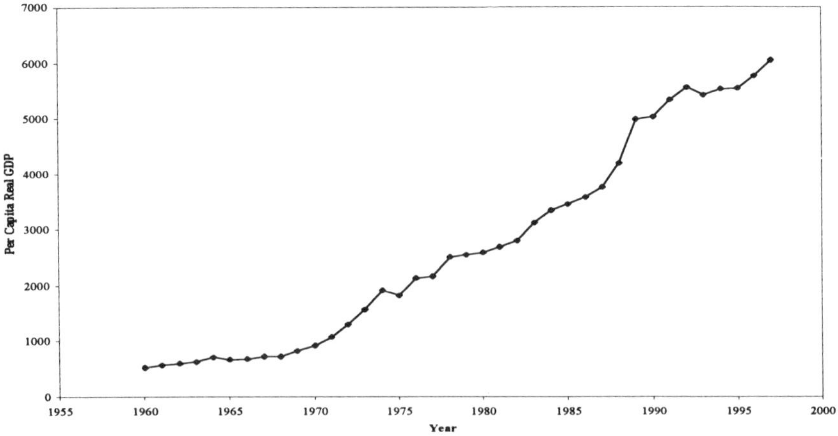

investment projects (three diamond mines were opened between 1971 and 1982); (2) improvement of the infrastructure; and (3) social investment, including in the areas of education and health, earning the country a high mark on the United Nations Human Development Index scale.

Careful management of public resources and public funds by the authorities has put the country in a strong fiscal position both internally and externally. Since fiscal year 1983–84 the central government budget has consistently been in the black; international reserves have steadily accumulated since 1976 to just over $6 billion in 1999, enough to finance imports for 29 months. Also, the country's current account and trade balances have remained positive. Other remarkable aspects of this economy is that continued growth has taken place while inflation has remained steady and has even decelerated since 1993; and, the government has kept external debt to low levels. The ratio of external debt to exports of goods and services was 1.3 in 1996, down from 5.5 in 1990. This stable and credible environment has provided incentives to the private sector to save and invest in the domestic economy. In 1998, for example, gross domestic savings as a percentage of GDP was 38.5 percent of which the private sector contributed 35.5 percent. In the same year, gross domestic investment as a percentage of GDP was 31.7 percent of which 19.2 percent was contributed by the private sector.

Prospects for Sustainable Economic Growth

As observed above, Botswana's economy is vulnerable to the vagaries of external economic forces due to its excessive reliance on the diamond sector. As an example of this dependence, in 1998 Botswana's diamond exports fell by 30 percent as a result of a purchasing quota imposed by the diamond-marketing cartel, De Beers. The fall contributed significantly to a slow down of the economy to an estimated 4 percent in 1998–99. The economy is also prone to the negative impact of drought on agricultural production. In the 1991–92 period, a combination of a weak diamond market, the most severe drought in the century,

Figure 7.2
Growth in GDP and Exports in Botswana Since 1960

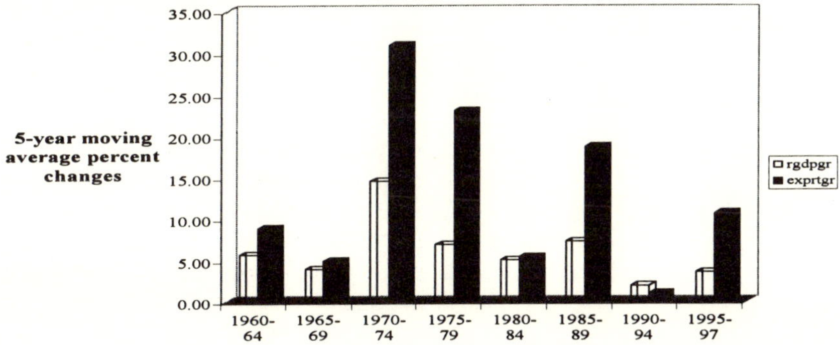

and financial irregularities mostly felt in the construction sector, pushed the country into recession. In 1993, the annual percentage change in real GDP was, for the first time, negative at –0.1 percent.

In response to these adverse situations, the authorities have made diversification and economic reform a priority in their development planning. The theme of the Eighth National Development Plan is sustainable economic diversification. The plan seeks to strengthen financial policies, introduce institutional reforms, and remove market impediments such as subsidies in order to foster private sector growth. External current account and capital account transactions have been liberalized, contributing to the growth of non-traditional exports such as vehicle assembly, textiles, and food processing. Overall, the plan commits the authorities to "continue prudent macroeconomic policies and promote an environment conducive to economic diversification, employment creation, and poverty alleviation" (IMF Survey, March 1997). On the basis of the continued commitment of the Botswana authorities to improving the welfare of all and the positive response of the private sector, prospects for sustainable growth in the new century are very bright for the country.

MAURITIUS

The Country, the People, the Land, and Its Resources

Mauritius is an archipelago located east of Madagascar in the Western Indian Ocean. The country includes the islands of Mauritius (largest), Rodrigues, Agalega, and the Cargados Carajos Shoals. Its total area is 2,040 sq. km, the equivalent of 787 sq. miles. The population remained under one million until 1987; now it is estimated to be about 1.13 million. The major ethnic group is formed by the descendants of Indian immigrants, who constitute more than two-thirds of the population. Other groups include the Creoles (people of mixed African and European descent), and Chinese and European minorities. The country's religions include Hinduism, Islam, and Roman Catholicism.

Mauritius is of volcanic origin. It is mountainous, particularly in the South, and is almost entirely surrounded by coral reefs. The climate is tropical and humid. Average annual precipitation ranges from 40 inches on the coast to about 200 inches in the plateau region. The island boasts of no natural resources except for its fertile soil.

The Government

Mauritius is a former British colony that achieved independence on March 12, 1968. From 1968 to 1991, the country was a constitutional monarchy, whereby the British queen, as represented by a governor-general, retained executive power, although only nominally. In March 1992, Mauritius became a republic. It has since then been ruled by a National Assembly, consisting of 62 elected representatives and 8 others appointed to ensure representation of various ethnic groups; a president, elected by the National Assembly as head of state, and a prime minister, appointed by the president as the chief executive.

Economic Growth and Policy Since Independence

In view of the high population growth and unemployment in Mauritius, James Meade, a Nobel laureate in economics, said some 30 years ago that "the outlook for peaceful development [in Mauritius] is poor" (Bräutigam, 1998). Indeed, when the country gained independence in 1968, its real per capita income, measured in 1970 prices, was just $300, unemployment was 20 percent and the population growth was 4 percent. The agriculture-based economy was almost entirely dependent on sugar production, which accounted for 20 percent of the GDP and provided over 50 percent of exports earnings. By 1993 though, the country had managed to increase its per capita income more than tenfold to $3,030. In 1998, per capita income has climbed to $3,685 moving from being classified as a low-income country to the class of upper middle-income countries alongside such countries as Malaysia, South Korea, South Africa, and Chile. This African lion has been roaring right alongside the Asian tigers. Growth has consistently remained around 5 percent to 6 percent annually since 1970, despite occasional declines in GDP (Figure 7.3 and Table 7.3).

Similar to its Asian counterparts, Mauritius' rapid and spectacular growth performance, first spurred by sugar production in the early 1970s, became strengthened when the country made the decision to establish an export-processing zone (EPZ) in 1976 to facilitate an export-led economic growth.

Since then the economy has gone through remarkable structural transformation. Diversification into manufacturing and tourism was very successful. Whereas sugar production accounted for 20 percent of GDP and over 50 percent of exports earnings in 1968, by 1993, the importance of sugar has been reduced to just 6 percent of GDP and 23 percent of exports earnings. In addition to sugar, the country is now a leading exporter of woolen knitwear. It also exports flowers, optical goods, jewelry, and toys.

The structure of the economy today is quite different from what it was at the time of independence. Agriculture went from being the main economic activity to accounting for 10 percent of the economy. Manufacturing provides 24 percent of the GDP; production in the EPZ accounts for 12 percent of GDP and almost 56 percent of gross exports earnings. Tourism has expanded from contributing

Figure 7.3
Real GDP Growth in Mauritius Since 1970

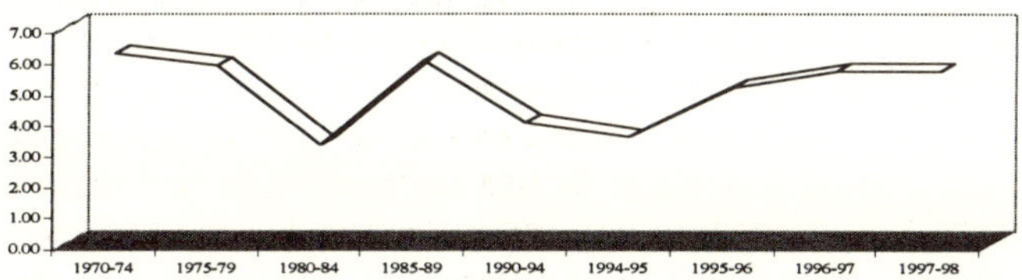

Table 7.3
Mauritius' Other Main Indicators

	1985	1990	1992	1993	1994
Gross domestic investment/GDP	23.5	30.4	28.6	30.5	31.7
Gross domestic savings/GDP	21.6	23.0	25.2	24.2	23.3
Exports/GDP	53.5	67.4	62.1	61.0	60.9
Total debt/GDP	58.4	37.9	34.1	30.4	31.0
Total debt/exports	108.8	54.4	52.5	48.1	49.0

Source: World Bank, *Trends in Developing Economies*, 1995

than 3 percent of foreign earnings in the 1970s to 19 percent in 1993. As a whole the service sector (including tourism, offshore banking and non-banking services, and the free port) constitutes 66 percent of the economy today. This is comparable to Singapore's.

Government Policy and Ethnic Relations

Much like the case of Botswana, the economic success of Mauritius also occurred within the framework of prudent and socially responsible utilization of the public resources by the authorities. The different governments have maintained a liberal economic policy while providing strong social protections, including social security, and free access to health care and education (Bräutigam, 1998). The country does not have a standing army. The savings from having no military expenditures were used to make great achievements in the indicators of human development, such as access to safe water guaranteed for 100 percent of the population, life expectancy rivaling any in the West at 70.5 years, and virtual elimination of illiteracy. The country is ranked 59 according to the 1999 United Nations Human Development Index ranking, placing it in the same category as Malaysia, Hungary, and Mexico. Reliance on external financing has greatly been reduced. The ratio of debt service to exports revenues has declined from 24.3 percent in 1985 to 6.7 percent in 1998, and the country has kept enough international reserves to cover at least 6 months' worth of imports.

With the exception of the communal riots in 1968 (certainly fueled by independence fervor) and riots in February 1999 (sparked by suspicions over the death of a popular singer), the island has proved James Meade wrong by not only achieving high levels of economic development, but by doing so in stability and peace. This is an important point to make here since the standard analysis coming from economic models seems to be that multi-ethnicity is a negative factor in development. In fact, for Mauritius ethnic diversity has been an engine of growth rather than an impediment (World Bank, 1995a). Further, the government has made tremendous efforts to ensure that the benefits of economic development accrue to all groups rather than a chosen few (although social problems of a racial nature continue to exist). A country's Gini coefficient gives a measure of the level of income inequality in that country. Mauritius' Gini coefficient has declined from 0.42 in 1975 to 0.38 in 1992. This reflects its increasingly even distribution of income. There are still major issues of distribution that must be resolved. A majority of the Afro-Creole group and of the rural Hindu dwellers still feel disadvantaged. This will be one of the challenges that the country will have to overcome.

Challenges and Prospects for Sustainable Economic Growth

The main challenge faced by the Mauritius economy today concerns its adaptability to globalization. The country has relied on preferential treatment of its products (traditionally sugar, but also later clothing) in European markets made possible by the guaranteed quota it was granted under the Lomé Convention's 1975 sugar protocol. When this preferential treatment comes to an end in an increasingly competitive global economy, Mauritius will have to compete with other low-cost countries in the areas in which it now holds comparative advantage. This may put the economy in a jam in view of the fact that labor cost is rising on the island. Indeed, the government has allowed firms to import low-wage labor from other countries, mainly China to compensate for the country's tightening labor market.

The country has also moved to enhance regional economic integration with other south eastern African countries both in trade and financial services. Although the economy is much more diversified than say, Botswana, efforts need to be continued in this area so as to be flexible and adaptable to the changing world economy. The remaining problems of income distribution must be addressed so that no group feels left out of the economic boom and events such as the riots of 1999 (which took on racial and ethnic overtones although they did not start that way) can be avoided in the future.

In all, Mauritius is well positioned to face these challenges. Judging from its continued expansion in all components of GDP in 1998–1999 (when GDP grew by over 5 percent while other countries were just recovering from the Asian and other crises), the country's prospects for continued and sustained growth are also very bright.

BENIN

The Country, the People, the Land, and Its Resources

Benin is a small West African country with a population of about five million, growing at a rate of 3.3 percent annually. The population consists of some 42 ethnic groups. The largest groups in the South are the Fon (and related groups) and the Yoruba; in the North the Bariba and Somba predominate. The country covers an area of 112,622 sq. km, the equivalent of about 43,483 sq. mi.

Benin's climate ranges from equatorial in the South to an increasingly arid and tropical in the North. Annual rainfall is about 35 inches. The country is bordered by Burkina Faso and Niger to the north, by Nigeria to the east, and Togo on the west side. It is endowed with fertile land mostly in the South, and a network of rivers and lakes that can be exploited for irrigation. Mineral resources include iron ore, phosphate, chromium, clay, marble, limestone, and petroleum.

The Government

Contrary to the cases of Botswana and Mauritius, the political history of Benin since independence has been either turbulent and unstable, or stable under guns. A democratic system was put into place upon independence on August 1, 1960 from France. However, just three years later, the first of many coups d'etat would occur. Between 1960 and 1972, the country was ruled by twelve different regimes, some lasting as briefly as six months to a year. Because of this political instability, there was no clear path of economic development pursued by the country during this period.

In 1972, a military and leftist regime took over the reins of power. Between then and 1990, the country embarked on a socialistic development based on a Marxist-Leninist philosophy. Legislative elections were held from 1979 on but they were meaningless since they were within the context of a one-party system. In 1990, a civilian-led national conference instituted a multiparty democratic system. A new constitution was approved by popular referendum. Since then, the country has been an example of political stability on the continent.

Economic Growth and Policy Since Independence

The Beninese economy is highly open, but it is considerably dependent on primary and tertiary activities. The primary sector accounts for about 33 percent of GDP and provides the main export commodity, cotton. Re-export activities provide about 45 per cent of export revenues. The economy is dominated by the tertiary sector (mostly commerce), which accounts for 53 percent of GDP. The secondary sector contributes just 14 per cent to the GDP. Benin is a member of the African Financial Community currency zone, along with thirteen other French-speaking African countries. The zone maintains a common currency, the CFA Franc. In 1994, after having maintained a fixed par value of CFAF50 to French Franc (FF)1 since 1949, the CFA was devalued to CFAF100 to FF1. This action was considered necessary to stimulate the competitiveness of mem-

Figure 7.4
Real GDP Growth in Benin Since 1965

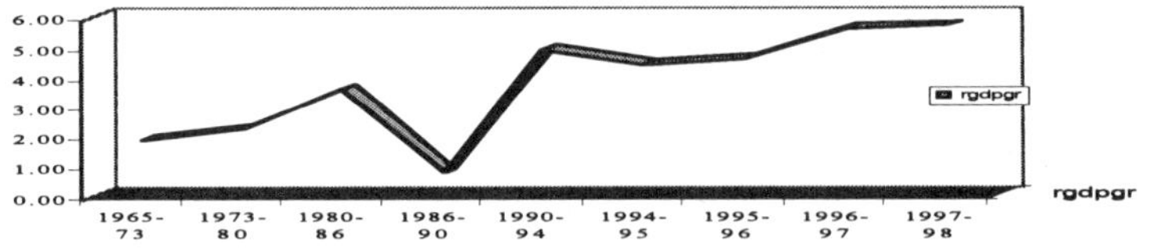

ber countries.

In contrast to the two countries already analyzed, economic progress in Benin was more or less a recent phenomenon. Average growth rates ranged from 0.52 percent to 3.7 percent throughout the 1970s and 1980s. The economy was actually regressing under the militaro-Marxist regime as evidenced by negative growth rates from the late 1970s and particularly throughout the 1980s. But since 1990, the country has seen a dramatic turnaround in its economic performance. The growth of GDP, which stood at a dismal -2.8 percent in 1989, rebounded to 3.2 percent in 1990, and has kept up rates of 4, 5, and 6 percent annually since then, making the country one of the fastest growing nations on the continent (Figure 7.4). A strong showing in the cotton sector (the main export product of the country), recovery of the textile sector (supported by new exports opportunities), and higher construction activities helped spur this strong growth performance (Table 7.4). In 1994, owing to sensible fiscal policy the primary current budget surplus rose to 2.8 percent of GDP. Increasing exports revenues and declining imports (due to the devaluation of the CFA Franc) resulted in the decrease in the current-account deficit from 7.5 percent of GDP in 1993 to 3 percent in 1994. This positive result in the current-account balance together with the surge in private capital inflows led to a surplus in the overall balance of payments estimated at $78 million in 1994 (World Bank, 1995a).

There are indications that the economic recovery is being felt at the micro level. Even though the state continues to exercise tremendous influence in the cotton sector, the advent of rural trade cooperatives has made it possible for the income of rural dwellers to increase with the higher cotton revenues. Also, after stagnating or even regressing in the 1980s, per capita real GDP has grown at 1 percent to 2.5 percent annually since 1991. The country has a long way to go to achieve the levels of human development observed in the other two countries considered in this chapter. However, if the government maintains its objective of "restoring a high and sustained rate of growth, improving living standards, and reducing domestic and external imbalances," there is no question that the country will rapidly achieve higher levels of human development.

Table 7.4
Benin's Other Main Indicators

	1985	1990	1992	1993	1994	1995	1996
Domestic investment/GDP	8.9	14.2	13.8	14.3	15.8	19.6	17.1
Domestic savings/GDP	0.8	5.5	3.9	3.5	9.5	10.2	8.7
Exports/GDP	34.4	21.8	23.3	22.1	26.9	20.0	19.0
Total debt/GDP	78.1	66.4	63.3	70.0	90.6	82.0	70.2

Sources: World Bank, International Monetary Fund, Government Statistics.

Challenges and Prospects for Sustainable Economic Growth

Benin's economic performance in the 1990s came about through the confluence of two major developments: (1) the country has finally settled its political mess, opting for a stable democratic system; and (2) the adoption of IMF/World Bank structural adjustment programs, albeit with their baggage of conditionality, also led to increased inflow of funds which the authorities have used wisely. Recent developments in the country have proven that it is possible to achieve enviable economic performance, even with low levels of resources, so long as there is stability and credibility. But Benin's external situation is cause for concern. Its dependence on mainly one export product, cotton; its long-term debt to exports ratio at 365 percent; and the debt service to exports ratio at 18 percent, constitute factors that expose this small country to the vagaries of the world economy. Diversification of the production base is absolutely necessary.

With its fertile soil and the diversity of crops including palm oil and fruits (some of the best-tasting pineapples are found in Benin) that it can produce, the country has no dearth of directions in which agricultural production can be diversified. Other sectors must be strengthened such as the textile sector and the mining and transparent marketing of the resources of the country. Private investment must be encouraged through gradual deregulation policies, removal of state monopolies, price controls, and other bottlenecks to investment where they exist. Continued restructuring of the banking sector and the strengthening of the regulatory and judicial environment will help attract private investment from within and from abroad. More important, continued improvement in infrastructure including roads and communication systems and a higher level of social spending including on health and education are factors without which the economic recovery will prove short-lived. Finally, attention must be paid to the issue of equitable distribution of resources across regions and across communities in this multiethnic country to insure that growth takes place in a peaceful environment (Cellule D'Analyse de Politique Economique, 1998).

The Beninese authorities are ready to meet these challenges. In section III of the Letter of Intent submitted to the International Monetary Fund in December 1998, the government declares that it intends to, inter alia, further reduce financial imbalances, keeping the economy on the path of sustained growth and job creation, and diversifying the production base; maintain a high GDP growth through higher levels of public investment, with priority given to investment in education, health, and infrastructure; increase savings and reduce the external debt burden. The authorities have also begun to implement the requirements of Articles 150–153 of the Constitution, which call for the devolution of power from the central government to the local communities, thereby ensuring increased control of the local communities over the management of their resources. This process, called decentralization, will surely help to enhance the feeling of ownership on the part of the different groups of the growth and development process underway in the country (Boko, forthcoming).

Overall, of the three countries examined in this chapter, the Beninese economy is the most fragile and delicate. It is poised to take off if the right environment (high infrastructure and social investment, credible financial and fiscal policy, etc.) continues to be fostered (Ndulu & O'Connell, 1999). It might falter if cautious management of the resources is not maintained. The authorities are aware of this delicate policy balance and have shown the willingness to focus their efforts and energy in the continued improvement of economic performance and social welfare. This bodes well for the future prospects of sustained growth and development.

CONCLUSION

This chapter attempts to demonstrate that despite the news of wars and other conflicts, tremendous economic progress has been achieved in post-colonial Africa, especially over the last decade. The three cases studied here represent only three examples of the many countries in which economic growth and development is being achieved. It should be reiterated, however, that the point of this chapter is not to suggest that post-colonial Africa has been an economic success story over all. Rather, it is to note that, contrary to the orthodox view, there have been remarkable achievements at different times and by different countries.

There are three lessons to be learned from this analysis: (1) a stable, predictable, and credible policy environment is necessary for sustainable economic growth, since it allows the private sector to make rational planning with no fear, for example, of their assets being ceased arbitrarily; (2) a cautious and transparent management and usage of a country's resources is growth-enhancing; it signals a sense of fair-play on the part of the authorities and it reinforces the credibility of announced policies; and (3) authorities should be flexible enough to adopt reform and restructuring, but in a prudent and socially responsible fashion. These lessons are being learned and being practiced throughout Africa today, and this speaks well for the future of the continent.

NOTE

I wish to thank Jonathan Mandelbaum and Chiara Gratton-Lavoie for valuable comments on an earlier draft. All remaining errors are mine alone.

REFERENCES

Barro, R. (1991). Economic growth in a cross section of countries. *Quarterly Journal of Economics, 106*, 407–443.

—— (1996). Democracy and growth. *Journal of Economic Growth, 1*(1), 1–27.

—— (1997). Determinants of economic growth: A cross-country empirical study. Cambridge, MA: MIT Press.

Benin Ministry of Economy and Finance. (1998, December). *Memorandum on Economic and Financial Policies for October 1998–September 1999*. Cotonou, Benin: Ministry of Information.

Benin Ministry of Finance/UNDP. (1998, July). *Bilan et Perspectives à Court et à Moyen Termes de l'Economie Nationale: Reformes Fiscales et Financement du Déficit Public*. Cotonou, Benin: Ministry of Finance.

Boko, H. S. (forthcoming). Ethnic diversity and economic growth: The search for stability in Benin. In P. Agbese (Ed.) *Government policies, the constitution and ethnic relations in Africa*.

Bräutigam, D. (1998, May). Economic takeoff in Africa. *Current History*, pp. 204–208.

—— (1999, May) Mauritius: Rethinking the miracle, *Current History*, pp. 228–231.

Cellule D'Analyse de Politique Economique. (1998). *Politique Macroeconomic au Bénin: Progrès, Limites et Perspectives*. Cotonou, Benin: Ministry of Planning.

Collier, P. & Gunning, J.W. (1999). Explaining African economic performance. *Journal of Economic Literature*, *37*(1), 64–111.

Easterly, W. & Levine, R. (1997). Africa's growth tragedy: Policies and ethnic divisions. *Quarterly Journal of Economics*, *112*, 1203–1250.

International Monetary Fund. (1995, December). Botswana accepts Article VIII obligations. Press release No. 95/63.

—— (1998, September). IMF Financial Statistics.

—— (1998, December). Benin: Enhanced structural adjustment facility policy framework paper, 1998–2001. Prepared by the Beninese authorities in cooperation with the staffs of the World Bank and the International Monetary Fund.

—— (1999, January). IMF approves second annual ESAF loan for Benin. Press Release No. 99/2.

—— (1999, August). IMF concludes Article IV consultation with Mauritius. Public Information Notice (PIN) No. 99/78.

—— (1999, November). IMF concludes Article IV consultation with Botswana. Public Information Notice (PIN) No. 99/105.

Landes, D. S. (1998). *The wealth and poverty of nations: Why some are so rich and some so poor*. New York: W.W. Norton & Co.

Microsoft. (1996). Encarta Encyclopedia: Benin, Botswana, Mauritius. Redmond, Washington: Microsoft Corporation.

Ndulu, B. & O'Connell, S. A.(1999). Governance and growth in sub-Saharan Africa. *Journal of Economic Perspectives 13*(3), 41–66.

Sachs, J. D. & Warner, M. (1997). Sources of slow growth in African economies. *Journal of African Economies*, *6*, 335–376.

Sender, J. (1999). Africa's economic performance: Limitations of the current consensus. *Journal of Economic Perspectives*, *13*(3), 89–114.

United Nations Development Program (1999). *Human Development Report 1999.* New York: UNDP.

Vaughan, O. (forthcoming). Chieftancy politics, communal cohesion and democracy in Botswana. In P. Agbese (Ed.), *Government policies, the constitution and ethnic relations in Africa.*

World Bank. (1988). *World Development Report 1988*. Washington, DC: World Bank.

—— (1989). *Sub-Saharan Africa: from crisis to sustainable growth.* Washington, DC: World Bank.

—— (1991). *World Development Report 1991*. Washington, DC: World Bank.

—— (1994). *World Development Report 1994*. Washington, DC: World Bank.

—— (1995a). *Trends in Developing Economies*. Washington, DC: World Bank.

—— (1995b). *World Development Report 1995*. Washington, DC: World Bank.

—— (1996). *World Development Report 1996*. Washington, DC: World Bank.

—— (1998/99). *World Development Report 1998/99*. Washington, DC: World Bank.

8

Developments in Transportation and Communications

Folu Folarin Ogundimu and Okechukwu C. Iheduru

This chapter discusses the development of Africa's transportation and communications infrastructure, showing the impact of colonial exploitation and underdevelopment. Furthermore, we intend to show that despite the shortcomings of Africa's misrule in the post-independence era, Africans have by and large done a commendable job in expanding access to the transportation and communications infrastructure bequeathed at independence.

African countries' policies of demonopolization of state control over wide areas of public economies provide additional examples of how post-independence Africa has come to grips with the realities of inefficient institutional models left in place from colonial times. These policies include the upgrading of public transportation and the modernization of communications infrastructure, most of which were established in the colonial era for the sole purpose of moving agricultural and mineral products from the extraction sites to the seaports. In the post-independence era such efforts focus largely, but not exclusively, on the provision of wider access to infrastructure, the revitalization of management, and the privatization of ownership. These reforms are concurrent with accelerated global transformations of institutional structures. Where reform was slow, it intensified in the aftermath of the collapse of the Soviet empire and the end of the cold war, partly because the field of maneuverability substantially narrowed for otherwise reform-resisting regimes.

As an analytic framework, we suggest that regardless of whether we consider the consequences of geopolitics or sectorial analysis, one implication of democratization in sub-Saharan Africa—essentially through pluralist choice policies—has been the emergence of wide-ranging institutional reforms in several policy subsectors such as those in the communications and transportation industries. These wide-ranging institutional reforms include denationalization, privatization, and implementation of free-market type policies under the general rubric of structural adjustment programs. For evidence, we rely solely on the record of achievements in publicly available documents and databases to examine what we consider an objective evaluation of the pre-independence and post-independence records of sub-Saharan Africa's infrastructural development.

Accuracy, comprehensiveness, and reliability of African databases are highly questionable. Nevertheless, we believe that few would dispute the general premises of the conclusions we offer in this chapter. Our brief discussion shows that the expansion of communications and transportation infrastructure in the post-colonial African state represents a tremendous leap of progress over what colonial administrations left in place at independence. Furthermore, attempts in recent years by African governments to implement institutional reforms while concurrently expanding modern infrastructure show that capacity building and infrastructural development are proceeding despite political, economic, and social difficulties that confront them domestically and globally. In the case of communications and transportation, one might argue that the devolution of state authority to subnational entities and to the private sector represents a further dismantling of, or modifications to, the last vestiges of inflexible command and control management structures inherited from colonial administrations.

Until the recent wave of reforms, many post-independence African governments had retained the colonial model of command and control governance and management partly because: (1) it was bequeathed to them at independence; (2) it was encouraged by international development agencies, such as the World Bank, who found such systems easily amenable to meeting result-oriented outcomes and bureaucratic accountability often demanded by the agencies, and (3) it provided the most opportunities for controlling the system of patronage or what others have termed "presidential monarchism," neopatrimonial state structures, or "prebendalism" (Fatton, Jr., 1992; Joseph, 1987). Aside from creating more equity-based, participatory, or egalitarian states, the policy reforms of the 1990s not only desired wider access to modern infrastructure, they encouraged more efficient maintenance and management cultures. Throughout this chapter, we refer largely to the experiences of Ghana and Nigeria as illustrations of how the record of colonial underdevelopment of sub-Saharan Africa's infrastructure is gradually, though painfully, being reversed in a more promising direction.

HISTORICITY OF INFRASTRUCTURE IN MODERN AFRICA

Few who know sub-Saharan Africa would argue that levels of efficiency in public transportation and communication are adequate. The evidence on this score is largely incontrovertible. Except for a few pockets in Africa's urban centers, communication and transportation services are either inefficient or inadequate. Data on the low density of transportation and communications infrastructure—compared to other regions of the world—partly illustrate the continent's low level of development. For example, in the cases of Ghana and Nigeria telephone density of three mainlines per 1,000 people by the mid 1990s was only slightly higher than the continent average of one mainline per 1,000 people. This means less than 1 percent of the population have access to telephones on the continent, less than one-half of the 2.8 percent average for developing countries (World Bank, 1994, Table 32). Another example of inadequate infrastructure in Africa is system losses from power outages in Nigeria and Ghana, which is 51 percent and 20 percent, respectively. Or to use the example of transportation, the World Bank reported in 1994 that one-third of Nigeria's roads and four-fifths of Ghana's roads were in poor condition (World Bank, 1994, Table 32).

Critics point to the inadequacy of public infrastructure development of these kinds as further evidence that post-independence transformations have failed to meet expectations of accelerated development in the wake of political emancipation. Accumulating evidence of mismanagement, corruption, and political tyranny by several African governments serve as ammunition for this criticism. However, no discussion of infrastructural development can be complete without close attention to the historical record.

First, one must acknowledge that the historical record on the colonial exploitation of African territories is straightforward. Transport and communications infrastructure were developed primarily to further colonial expansion, exploitation, and expropriation. According to Gordon (1992), regardless of the complexity and variegated consequences of colonial economic history, European strategies and actions were the same.

> African colonies were made politically and economically subordinate to European needs. The recently industrialized states of Europe, particularly Britain and France, required cheap raw materials and desired captive markets for manufactured goods. Over time, these governments integrated their African colonies into what many call the "international capitalist system." (Gordon and Gordon, 1992, p. 53)

Samir Amin's landmark study on the consequences of colonial domination was even more bitingly specific. Africa, Amin says, was a source of cheap raw material, primarily agricultural commodities such as rubber, cocoa, palm oil, groundnut, cotton, and such minerals as copper and gold necessary for newly industrializing Europe. Thus, pre-independence Africa was thoroughly integrated into an international market system dominated by European powers (Amin, 1972). Rodney's (1984) seminal work, *How Europe Underdeveloped Africa*, though impugned by defenders of the status quo, provides cogent evidence of colonial exploitation.

The cases of the development of transportation infrastructure in Ghana and Nigeria are perfect illustrations. Roads and railway links were built not for the purpose of servicing the interests of citizens of the colonial state but rather as conveyor belts linking arteries of production in the colonial hinterland with coastal ports for shipment of raw material to the metropole. Consequently, much of the rail infrastructure in Ghana and Nigeria—though primitive—was already set by independence. By independence in 1957, Ghana had just under 1,000 kilometers of rail link, much of that network built between 1920 and 1940, a period that saw a 45 percent growth in rail infrastructure. Between 1940 and 1957, there was a further 15 percent growth. After independence, there was about 0.1 percent growth in Ghana's rail infrastructure (Economic Commission for Africa (ECA), 1990; Mitchel, 1982).

Similarly, Nigeria's rail infrastructure grew by 41 percent between 1920 and 1940. After 1950, there was a slight decline from 3,063 kilometers to 2,864 kilometers. Further extension took place after independence in 1960, with a 19 percent expansion of the rail infrastructure to 3,523 kilometers.

If other transportation modes adjusted to changing circumstances, so did shipping and ports, even though there was a great time lag. The African coastline is today dotted with nearly 150 seaports of which some 60 could be classified as international. At independence, many African countries inherited seriously inadequate port facilities; it is only in the last 30 years that basic port infrastructure has been provided by the post-colonial states. Since 1960, the number of berths has increased from about 150 to over 600, and many existing ports were extended and improved (Port Sudan, Mombasa, Dar-es-Salaam, Pointe Noire, Douala, Lagos-Apapa, Takoradi, Abidjan, Freetown, and Dakar); new ports were built, some for general cargo (San Pedro, Tema, Lome, and Cotonou); and others for bulk minerals (Nouadhibou, Buchanan, and Kpeme). In oil exporting countries, numerous oil terminals were provided. All this has been achieved at a time when the technology of maritime transport was itself undergoing revolutionary changes.

Because most of the continent's coastline is inhospitable and natural ports are few in number, port development has often involved difficult engineering at great costs. Even though much of these great increases in infrastructure were not matched everywhere by optimum utilization and acceptable levels of productivity, credit must still be given to the post-colonial states for taking the first steps to launch these countries on the path to accelerated development (Hilling, 1990, pp. 330–335; Iheduru, 1995).

Overall, the commitment of the post-colonial state in providing transport and communications services has been tremendous. Some estimates show that 20–25 percent of national (limited) resources were invested in these fields from 1960 to 1988. At the beginning of the United Nations Transport and Communications Decade in Africa, 1978–1988 (UNTACDA), 771 projects costing $9 billion were identified. Of these, 610 costing $8.3 billion, were in the transport sector and the balance in the communications sector. It has also been shown that over 40 percent of all World Bank-financed transport projects, and 20 percent of the lending amounts involved, took place in sub-Saharan Africa in the 1978–1982 financial years (Economic Commission for Africa, 1979, pp. 2–3). These figures and the amounts involved are monumental if compared with the developmental efforts of the colonial governments in these sectors.

Despite the provision of colonial rail infrastructure, the system was essentially designed for what we already identified as colonial expansion, exploitation, and expropriation. In reference to the antiquated nature of colonial transport infrastructure, Davidson (1983, p. 200) has remarked that:

> Colonial Africa had only a primitive infrastructure of modern equipment. There was, for example, a shortage of modern sea-ports. Ghana had only Sekondi for the export of minerals and cash-crops; Nigeria had only Lagos. . . . But independence soon improved this situation. New ports were built at Tema in Ghana, at Abidjan in Ivory Coast, and elsewhere in the west. . . . Existing railways were improved and some new ones built.

Although colonial transportation was inadequate for system efficiency and coverage or accessibility to the majority of population, metropolitan powers nevertheless exploited this critical resource as a tool for the pacification and exploitation of the continent (Davidson, 1983; Neff, 1992). Ironically, much of the infrastructures in the rail and sea transport modes have continued to perpetuate these colonial development patterns.

Viewed as cultural or economic anthropomorphism, the impact of transportation on Africa is not unique in this respect. Bonnafous (1991) and other economists go as far as acknowledging that transformations in human civilization have often occurred because of the availability of transportation to move people and goods from one place to another. Consequently, immense changes in national economies are sometimes linked to developments in transportation because of the relative cheapness and speed of moving goods and people (Bonnafous, 1991, p. 83).

One example of how rail infrastructure was essential to the system of commerce, the movement of goods and of passengers, is illustrated by trends in freight and passenger traffic on Ghanaian and Nigerian railway systems. As indicated, much of the rail links in Ghana and Nigeria were already established at independence. For Ghana, independence came in 1957, while Nigeria's was in 1960. Except for improvements to rolling stock and upgrades of existing infrastructure, post-independence African governments did not extend these rail networks, focusing instead on improvements to existing rail links while developing access to alternative modes of transportation such as roads, waterways, and transport.

For example, the colonial government's extension of Ghana's rail network between 1925 and 1955 resulted in the tremendous growth of freight and passenger traffic. Volume of freight on railways grew by 63 percent during this period, reaching a peak of 2,208,000 metric tons by 1955. Passenger traffic also grew by about 70 percent during the same period. Similarly, rail freight traffic grew by 70 percent and ridership grew by 62 percent between 1925 and 1955 in Nigeria. However, both freight and passenger traffic declined substantially in the post-independence era, partly because of the development of alternative modes of transportation such as roads and waterways, and partly because of underfunding of the rail transportation system.

The declines mirrored similar downturns in rail transport all over the world with the ascendancy of the automobile and the availability of cheap oil from the Middle East and other oil-exporting countries. Development agencies also began to de-emphasize the funding of rail transport projects as a reflection of the political and economic trends in the industrialized countries. Moreover, overseas demand for many of Africa's mineral and agricultural exports has been either stagnant or expanding very slowly, partly because technological innovations in the industrial countries have led to both the substitution of certain raw materials and conservation in their use (Mikesell, 1988, pp. 41–47). In 1982, Kamarck (1982, pp. 149–163) forecast that there would be "no major increase in the exploitation of non-fuel mineral resources of most of tropical Africa before the year 2000." As export-freight requirements remained at a modest level, the state had very little incentive to expand access in these transportation modes (Moriarty, 1989, p. 131).

In Ghana, all the factors discussed above resulted in a dramatic decline in freight traffic by 36 percent by 1970, and by 93 percent between 1970 and 1987.

Loss in ridership for Ghana's railway system was 53 percent between 1960 and 1985. Loss of freight and passenger traffic for the Nigerian system was less dramatic but nevertheless substantial. The system lost 65 percent of its freight traffic between 1955 and 1970 but witnessed a modest growth of 6 percent between 1970 and 1987. In the decade after independence, the system lost about one-fourth of its passenger traffic. Losses in freight and passenger traffic on the rail system were more than offset by developments of alternative modes of transportation as previously mentioned.

It should be noted in passing that despite criticism of poor road networks in Ghana and Nigeria, the two governments have spent considerable sums of money expanding road links with previously inaccessible parts of their countries. For example, in Nigeria the road network was expanded by a further 5 percent between 1978 and 1987, to some 110,300 kilometers. During the same period, Ghana expanded its own road network by 12 percent, to 35,740 kilometers. The growth in commercial and private vehicular traffic in post-independence era is one measure of how fast access to private and commercial transportation have grown in the post-colonial state. In Ghana, this growth was 65 percent between 1958 and 1975 for commercial traffic. In Nigeria, the growth was 41 percent between 1960 and 1985.

In Ghana, air traffic for passengers and cargo grew by over 90 percent between 1960 and 1975, and motor vehicles in use grew by 80 percent between 1958 and 1991. Similarly, in Nigeria air traffic grew by 90 percent and motor vehicles in use grew by 86 percent between 1960 and 1987. Developments in shipping more or less parallel growth in other sectors of the transportation industry.

As with transportation, communication facilities are generally not well developed in sub-Saharan Africa, especially when compared with the level of development in other parts of the world. Again, critics are neither sympathetic nor impressed with why the low-level development of communications infrastructure persists after more than 30 years of independence. Earlier, we referred to the poor density of telephones on the continent. Poverty of communication media is not, however, limited to the telephone. Despite a long tradition of the mass circulating press and increases in the diffusion of hardware to facilitate greater access to mass communication, Africa remains the world's least developed region in the circulation of newspapers, and in the availability of radio and television receivers. The reasons for the lack of progress in this area are many and complex, but three of them stand out: (1) the specter of authoritarian African governments denying Africa's press a vital role in public affairs;(2) lack of a broad support base, with the media serving the interests of a narrow elite rather than as a mass medium, and (3) absence of favorable economic and social conditions that would ensure survival of media as commercial enterprises (Hachten, 1993).

Valid as these explanations are, they nonetheless ignore two other interrelated factors. First, they ignore the colonial origins of the repression of media. Second, they ignore the role played by strong economic environments in the emergence of free-market, competitive, private, and commercially-viable media and communications enterprises. Experience from other parts of the world, including those of Asia, the United States, Eastern and Western Europe show that without viable, competitive market systems and without the appropriate balancing of regulatory environments through careful public policy, viable private media and communications enterprises do not emerge. This leads us to a consideration of the policy environment in sub-Saharan Africa and the impact reformist policies are having for subsectors such as transport and communication.

REFORM AND SUBSECTOR DEVELOPMENT

Much of the reform effort in sub-Saharan Africa's subsectors come under policies broadly linked with much-maligned structural adjustment programs (SAPs). SAPs are euphemisms for a wide range of World Bank and International Monetary Fund (IMF)-inspired macroeconomic and institutional reforms. The long-range objectives of the programs include cutting government expenditures, enhancing revenues, and moving toward market-based strategies in exchange rates, prices, import and export controls as well as the divestiture of state-owned enterprises (SOEs).

In the case of sub-Saharan Africa, the success of these programs has been mixed (Jones & Kiguel, 1994; Wilson, 1992, March 18). SAPs, many believe, have been less successful than the international financial institutions are willing to concede. Ayittey (1992, p. 309) dismisses with disdain Africa's attempts at structural reform in the 1980s, arguing that in most cases "structural adjustment amounts to reorganizing a bankrupt company and placing it together with the massive infusion of new capital, in the hands of the same incompetent management that ruined it in the first place." Whereas the World Bank is generally enthusiastic about the institutional outcomes of structural reform programs, others have focused on the social outcomes of adjustment policies as well as the failure of SAPs to invest decision making capabilities and flexibility in Africans themselves (Economic Commission for Africa, 1989; Fatton, 1992). According to Fatton (1992, p. 127), "there is little evidence that structural adjustment programs have contributed to economic recovery. On the contrary, they have exacerbated social conditions and inflicted severe losses on the poor."

Leaving the social consequences of adjustment programs aside for a moment, one might ask two related questions in respect of subsector development in sub-Saharan Africa. First, what kinds of investments in common-carrier infrastructural projects such as transportation and communication have been made by African governments? Second, what kinds of commitments to institutional reform efforts are evident for continuing subsector development? These two questions are appropriate because reform-inducing agencies such as the World Bank now say infrastructural investments, maintenance of old infrastructure, and implementation of more efficient delivery systems are critical for bringing about rapid development (World Bank, 1994). As for the first question, the Bank estimates that developing countries as a whole would have to invest $200 billion a year—4 percent of their output, and one-fifth of total investment—in infrastructure services such as transport, power, water, sanitation, telecommunications, and irrigation.

Given Africa's current dependence on the international community for the financing of major development projects, it is conceivable that these massive infrastructural investment programs will be a priority focus of the World Bank and other international development lending agencies for the next several years, provided the recipient governments are democratic. This has been the case in the recent past. For example, Ghana attracted about $3 billion in overseas development assistance between 1985 and 1991 alone, while Nigeria attracted $2.1 billion (World Bank, 1994, Table 19). Much of these receipts went into funding public infrastructure projects.

Recently, some African governments have made significant investments in public infrastructure programs. Ghana invested 10.5 percent of total expenditure in transport and communications in 1979. In 1988, the share of total expenditure budget spent on transport and communications was 6.1 percent. Similarly, Nigeria has increased its investments in transport and communications over the last several years. In 1992, transport and communications accounted for 22 percent of the capital budget and 17 percent of the recurrent budget. In its medium-term rolling plan (1992–1994), transport and communications accounted for less than 1 percent of planned expenditure, but its budget was 1.5 billion naira. At the time of release of the rolling plan, the Nigerian government said full implementation of projects in the plan would boost the growth of the Gross Domestic Product by an optimistic 6 percent. Whether or not such optimism was justified is a different story. But judging by the record of recent contributions of transport and communications to GDP, it is not unreasonable to argue that such public infrastructural development projects hold promise for upgrading Africa's development objectives. For example, despite negative growth rates of the Gross National Product (GNP) per capita throughout the 1980s in Ghana and Nigeria (World Bank, 1994), the growth in the contribution of transport and communications to GDP was 94 percent for Ghana and 58 percent for Nigeria between 1978 and 1987.

Aside from the not-too-shabby record suggested by these macroeconomic indicators, the emergence of commercial, privately-owned communications enterprises is perhaps the brightest development in the communications subsector in sub-Saharan Africa. As the pace of democratization quickens on the subcontinent, many countries are finally living up to newly-enshrined constitutional promises that allow for the emergence of privately-owned media and communications companies. These new companies operate side by side and compete with state-owned enterprises. Nigeria is providing significant leadership in this area, while Ghana is not too far behind in implementing similar programs. At the time of this writing, Nigeria's National Communications Commission (NNC) had already issued licenses to 114 major private commercial operators of radio, television, and satellite retransmission services. Twenty-four of them had already gone on the air, providing a fare of programming many previously thought unthinkable a few years ago.

Furthermore, critics of Africa's post-colonial record gloss over many of the current transformations taking place in Africa's communications sector such as the emergence of the Regional African Satellite Communications (RASCOM) project. The goal of the project is to establish a pan-African satellite system for network linkage with all African countries. Also, there is the Pan African Development Information System (PADIS), which is charged with improving information infrastructure for the creation of databases in fields of importance to development in the region. The revitalization of the Pan African News Agency (PANA) is leading to the strengthening of open, pluralist media systems throughout the continent by encouraging the dissemination of news and information of all kinds that Africa's media systems would find essential for their own various uses. One might argue that these attempts to formulate common standards and encourage the emergence of common carrier networks linking every part of the continent represent one of the most important achievements of the post-independence African state.

Substantial progress in creating necessary institutional capability as well as infrastructure has been coupled, moreover, with pressure to reform existing national communications environments. It is precisely such changes in regulatory environments that have led to the emergence of pluralist, dynamic, and abundant media in some countries. The sense of optimism that is greeting these developments was properly noted by four Africanist scholars who found corroboration in the words of a United Nations Africa observer: "Africa is not all gloom and doom: The 1990s saw important developments toward democracy and the strengthening of the media" (de Beer, Kasoma, Megwa, & Steyn, 1995, p. 211).

SUMMARY AND CONCLUSIONS

In this chapter, we have argued that there is a lot to celebrate in Africa's record of post-colonial achievement in the transportation and communications sector. Without historical hindsight, it is easy to dismiss the present inadequacy and, sometimes under-utilization, of Africa's infrastructure as a consequence of bad governance. On the contrary, we argue that such conclusions suffer from historical amnesia. We argue that the African state, first and foremost, was dealt a bad hand at independence. In the subsector of transport and communications, we have shown in this chapter that at independence Africans inherited antiquated infrastructure, managed by centralized, authoritarian systems designed primarily to service the interests of metropolitan powers in Europe.

As bad as the record of post-independence governance has been, two facts are indisputable from the evidence reviewed in this chapter. First, sub-Saharan African governments like Nigeria and Ghana have spent considerable portions of their limited resources in expanding access to modern infrastructure, including transportation and communications. These programs have facilitated the development and improvements in several modes of communication and transportation that were not provided during the colonial era. Second, the pressure to reform authoritarian governance structures are finally leading to the emergence of pluralist and commercially viable communications systems in many African states. One hopes that these new systems will provide and sustain the element of choice, promote accountability in governance, and encourage the emergence of affordable, efficient transportation and communications throughout the continent. Third, the focus on creating and strengthening Africa's management and institutional capacity should lead to the emergence of efficient and appropriate regulatory environments and policies under which vital communications projects will thrive in the new information age.

REFERENCES

Amin, S. (1972). Underdevelopment and development in black Africa: Origins and contemporary forms. *The Journal of Modern African Studies, 10*(4), 503–524.

Ayittey, G. (1992). *African Betrayed.* New York: St. Martin's.

Bonnafous, A. (1991). Transport problems and topics for research. *Impact of science on society, 4*(2), 83–88.

Browne, R. S. (1984). *The Lagos Plan of Action vs. the Berg Report.* Washington, DC: African Studies and Research Program, Howard University.

Davidson, B. (1983). *Modern Africa.* New York: Longman.

de Beer, A. S., Kasoma, F. P., Megwa, E. R. & Steyn, E. (1995), Sub-Saharan Africa. In J. C. Merrill (Ed.), *Global Journalism: Survey of International Communications* (3rd ed.) (pp. 209–268). New York: Longman.

Economic Commission for Africa. (1979). *United Nations Transport and Communications Decade in Africa, 1978–1988.* Addis Ababa: United Nations.

—— (1989). *Africa's alternative framework to structural adjustment programs for socio-economic recovery and transformation.* Addis Ababa: United Nations-Economic Commission for Africa.

—— (1990). *African statistical yearbook 1987* (Part 11). Addis Ababa: United Nations Economic Commission for Africa.

Fatton, R., Jr. (1992). *Predatory rule and civil society in Africa.* Boulder, CO: Lynne Rienner.

Gordon, D. L. (1992). African politics. In A. A. Gordon & D. L. Gordon (Eds.), *Understanding contemporary Africa* (pp. 51–85). Boulder, CO: Lynne Rienner.

Hachten, W. A. & Hachten, H. (1993). *The growth of media in the Third World: African failures, Asian successes.* Ames, IA: Iowa State University Press.

Hilling, D. (1990). African development and the maritime sector. *The Dock and Harbour Authority, 70*(April), 330–335.

Iheduru, O. C. (1995). *The political economy of international shipping in developing countries.* Newark, DE: University of Delaware Press.

Jones, C. & Kiguel, M, (1994). Africa's quest for prosperity: Has adjustment helped? *Finance & Development, 31*(2), 2–6.

Joseph, R. (1987). *Prebendal politics in Nigeria.* Cambridge: Cambridge University Press.

Kamarck, A. M. (1982, Spring). The resources of tropical Africa. *Daedalus*, pp. 149–163.

Mikesell, R. F. (1988). The changing demand for raw materials. *Economic Impact* (Washington, DC), *64*, 41–47.

Mitchel, B. R. (1982). *International Historical Statistics: Africa and Asia.* New York: New York University Press.

Moriarty, P. (1989). Transport in tropical Africa. *The Journal of Modern African Studies, 27*(1): 125–132.

Neff, J. W. (1992). Africa: A geographic preface. In A. A. Gordon & D. L. Gordon (Eds.), *Understanding contemporary Africa* (pp.7–20). Boulder, CO: Lynne Rienner.

Rodney, W. (1984). *How Europe underdeveloped Africa.* Washington, DC: Howard University Press.

Wilson, E. (1992, March 18). Testimony. Joint hearing before the subcommittee on foreign affairs, *House of Representatives,* 102nd (U.S.) Congress, second session.

World Bank. (1994). *World development report 1994.* Washington, DC: World Bank.

Part V

Health and Agricultural Innovations

9

Post-Colonial African Achievements in Health

Emmanuel U. Nnadozie

This chapter reviews the post-colonial achievements of African countries in health, healthcare resources, and healthcare management. The chapter reviews inter-temporal achievements in health since the 1960s, the sources of accomplishments, and inter-country performance. Achievement in health is generally measured in terms of traditional temporal performance indicators such as life expectancy, infant mortality, child mortality, maternal mortality, and so on. In addition to these indicators, this chapter examines other nontraditional indicators as well as the importance of traditional healers in post-colonial Africa.

The World Health Organization (WHO) defines health as "a state of complete physical and mental and social well being and not merely the absence of disease or infirmity." Since poor health reduces labor productivity, health plays a vital role in economic development. Together with adequate nutrition, primary education, sanitation, water supply, and housing, health is a basic human need. Lack of good healthcare especially for pregnant women and children leads to deficiencies in children's mental and physical development. Better health, therefore, brings substantial economic benefits such as enhanced labor productivity, which, in turn, contributes to the overall growth of the economy. Improved health contributes to economic growth by improving school enrollment and performance, and reducing costs of medical care (World Bank, 1993). Given the importance of good health in society, government plays a key role in compensating for problems generated by uncertainty and market failure. It can improve the economic environment for healthy households, invest in public and essential clinical services, and reform the health system to make it more efficient.

This chapter shows that there have been appreciable and recognizable achievements in health in Africa since independence. It shows that in spite of well-known socioeconomic and epidemiological problems, the health of Africans improved after decolonization. Of course, the recent scourge of AIDS has negated some of the achievement. Yet an examination of the post-colonial record, shows that individual countries made considerable progress in health profiles and in human outcomes, expenditure on health and policy, and provision of resources. The important thing to note is that African countries' dynamic goals are to improve the human condition within their region and to change the structure of their human capital for long-run gains. African countries, therefore, aspire for health improvement as an approach to economic development rather than to increase aggregate welfare in the short run. Hence, results should be judged mainly by inter-temporal absolute changes in existing patterns of life expectancy, infant mortality, and female mortality, and so on, and their impact on socioeconomic and political problems. This calls to question the relevance of the existing orthodox theory of health behavior in Africa.

Notwithstanding the demonstrable accomplishments, the majority of literature tends to emphasize the problems and the failures of healthcare delivery in Africa. The vast amount of current literature on the AIDS crisis epitomizes this bias. Likewise, much of the information about African health conditions tend to use unreliable data to compare Africa with its former colonizers and other industrialized countries rather than highlighting accomplishments by individual countries. To effectively deal with post-colonial achievements in health, this chapter is divided into five sections. Section 1 gives a synopsis of the theoretical background, the conceptual framework, and pertinent definitions; section 2 gives a background of colonial neglect of healthcare and the selective provision of healthcare in Africa, to illustrate the enormous health difficulties inherited by African administrators at independence; section 3 deals with the general trend in major health variables in the continent such as life expectancy, infant mortality at birth, expenditure on health, and nutrition; section 4 discusses the accomplishments in terms of the components of a health system and gives a comparison among groups of countries—low-income, lower middle-income, and upper middle-income; and finally, section 5 elaborates on the role of traditional medicine, and is followed by discussions of the factors of success and failures.

This chapter adopts a regional approach, focusing on aggregate and individual country achievements, using the performance indicators of countries. Where possible, country-specific examples have been used. In the case where data are available only for sub-Saharan Africa, the analysis is done on that basis. The chapter does not undertake a detailed analysis of each African country's performance in health, nor does it undertake an analysis of the problems that may plague African heath care systems. Against this background, the impact of the AIDS pandemic is not discussed because it is not central to the thesis of this chapter. Likewise, because of its unique circumstances, healthcare in South Africa is deliberately ignored. One can, however, compare apartheid South Africa's health system to that of the colonial era because the apartheid health system was oppressive and access to health services was limited and in some cases nonexistent to Africans.

In general, data were sought for the period 1960–1993. For many of the requisite elements, data were available only for certain periods. Actual data from different sources show significant variation even when they describe the same conditions. Therefore, the data used in this chapter are subject to limitations due to their insufficiency and their lack of consistency when obtained from different sources. The majority of the data are from secondary sources including documents from the World Bank, the United Nations Development Program (UNDP), and the World Health Organization (WHO).

This chapter fills the gap in the existing literature as researchers have often studied the healthcare question as a problem not as an aspect of dynamic economies that evolve positively in certain aspects and negatively in others. Since the findings of these researchers may have been implicitly overgeneralized to all types of economies, this chapter augments existing literature by highlighting the achievements rather than the crisis. In spite of heavy reliance on existing data, this chapter recognizes the obvious limitations they pose. Likewise, the chapter recognizes the problem of attempting to generalize about health for a continent as culturally, politically, and economically diverse as Africa.

THEORETICAL BACKGROUND AND CONCEPTUAL FRAMEWORK

It is important to understand the theoretical framework on which the health demand function lies, so that we can adequately interpret data and results and make informed inferences. Do Africans see health as an end in itself or do they see it as a means to an end? For macro models, strong associations exist between life expectancy and per capita income. They also note the strong role of women's education in nutrient intake (Behrman & Deolalikar, 1988). The problem with applying the neoclassical demand theory to Africa is that at present, Africans do not ask for quantities of healthcare for the sake of being healthy, so they do not express their demand in terms of quantitative preferences subject to specified constraints. Rather, they ask for good health as a means of accomplishing more advanced socially defined goals such as raising children, providing for the family's security, and protecting the family. This behavior is not irrational even though it differs from the neoclassical norms existing in Western society.

If people believe that health lies in the hands of the divine power as Stock (1980) shows in the case of the Nigerian Hausas, they may not increase their demand for good health as their incomes rise. This does not and should not be interpreted as irrational behavior in Africa and other regions of the world where cultural beliefs and religion play a key role in the demand and supply of health services. This issue underscores the social and spiritual theories of disease causation as underlined by Twumasi (1982, p. 202), thus: "The traditional cosmology seemed opposed to scientific conceptions of disease."

Related to the inadequacy of the neoclassical model are the limitations posed by the use of conventional data in measuring health performance. For instance, such data exclude traditional health practitioners; they are also based on WHO standards of qualification to determine who is categorized as a physician. Therefore the local "chemists" and traditional birth attendants are not accounted for in spite of their significant contribution to healthcare. The same goes for measures of funding for hospitals and public healthcare, which often reflect the central government's allocations, ignoring often substantial local finances and other informal sources of funds.

Health is therefore determined by the quality of the physical and social environment, and individual traits. In addition to cultural variables, these three factors are enhanced by the availability of health services. However, the level of economic development affects supply and quality of health resources such as manpower, facilities, commodities, and knowledge (technology); it also affects medical education and the output of doctors. In addition to the aforementioned variables, other factors capable of affecting health in societies are urbanization, industrialization, education, government structure, international trade, and demographic changes. The extent to which these variables have been affected by colonial and post-colonial government policies explains contemporary health situations and determines the operation of the health system.

A health system consists of "the combination of resources, organization, financing, and management that culminate in the delivery of health services to the population" (Roemer, 1991, p. 31). A health system therefore encompasses the combined structure and functional relationships of the resources and actions. Since the primary objective of a health system is to promote health or recover it in case of disease or injury, it is pertinent to ask the question, what are the concrete accomplishments of African governments with regard to improving the major components of the health system, in terms of production and promotion of resources, organization of programs, economic support mechanism, management, and delivery of service? Before answering this question, it is important to examine the legacy of the colonial health system.

THE COLONIAL LEGACY

Though colonialism is not often thought of as a health hazard, several studies have demonstrated its deleterious impact on the health of Africans (Azevedo, 1988; Chuke, 1988; Coquery-Vidrovitch, 1989; Ityavyar, 1992; Mburu, 1992; Pearce, 1992; Shapiro, 1983; Twumasi, 1992). The findings indicate that African mortality levels rose as a result of the colonial conquest and attempts to dominate Africans during the late 1800s and the early 1900s.

It should be noted that some colonial governments established health infrastructure in some countries such as Nigeria and Kenya, but in other countries such as Chad, Mali, and Mozambique, health was neglected. There were also significant developments and improvements in the prevention of some communicable diseases during the colonial period. But, over all, colonial policies on health were selective and pragmatic. Medical services were geared primarily to protect white settlers, prolong their lives, and facilitate their colonial duties. The pattern of providing selective access to health and lack of development of health services was common among various European colonial governments. In Belgian Congo, for instance, the colonial administration initially concentrated on the establishment of hospitals for Europeans at the urban centers and later built a few hospitals for indigenes. Rural Congo was completely ignored.

There was also a serious problem with policy coordination. In Nigeria, for instance, despite the contributions of missionaries in the late 1800s and by the colonial administration early in the 1900s, health services remained uncoordinated with no coherent health policies or plans. These all changed significantly following independence.

Thirty years after Nigeria's independence in 1960, for instance, each of the country's 30 states had a Ministry of Health in addition to the federal Ministry of Health at the center. From one medical college, which later became the University of Ibadan medical school at independence, the country had over 13 medical schools by 1993. The nation's physician supply increased by over 60 percent from 1960 to 1979 as hundreds of hospitals sprung up across the country. Under French domination, Cameroon had limited health facilities as the following passage from Azevedo (1988, p. 123) illustrates:

> Prior to independence, there were 12 hospitals, 25 dispensaries, and 40 subsidiary health centers scattered throughout Cameroon. Each of the nation's five-year development plans since independence has included provisions for expanding health care facilities. Within a decade after gaining independence, facilities had expanded to 85 hospitals, 340 dispensaries, and 50 health centers as well as modernization of hospitals in Douala, Yaounde, Foumban, Garoua, and Ngaoundere and the establishment of the nation's only medical school at the university of Yaounde. By 1977, there was a combined total of 1,003 health care facilities.

As will be demonstrated below, the progress is reflected in various measures of health status and services.

POST-COLONIAL ACHIEVEMENTS

This section takes a three-dimensional view of post-colonial African achievement in health. First the health status in the continent and in African countries is evaluated in terms of the performance of the traditional health indicators such as life expectancy, mortality rates, population per physician and per nurse, daily per capita caloric supply, percentage of population with access to safe water, and the percentage of central government expenditure. Second, a systems approach is used to examine the delivery and provision of health by assessing the accomplishments in terms of the major components of the health system—resources, organization, economic support, health management, and improved health delivery. Finally, comparative analysis is used to show evidence of inter-group achievements. Where possible, specific country examples and studies are used to illustrate the significant accomplishment in health. Not only has the health status of Africans improved since independence, but the mobilization and use of resources have expanded. The organization, economic support for healthcare, health management, and health delivery have also improved.

The general picture of health and related indicators is not as gloomy as most analysts tend to portray. Total fertility rate, crude birthrate, infant mortality rate per 1,000 live births are all on the decline. In contrast, life expectancy and primary school enrollment are on the increase. In spite of innumerable constraints and problems, data from various sources confirm a significant level of achievements by African countries during the post-colonial period (Table 9.1).

Life Expectancy

Like the rest of the world, Africa showed significant improvements in life expectancy between 1950 and 1990, rising from about 38 to 52 years. Notably, significant accomplishments have been recorded in the 30 years leading up to 1990. As Table 9.1 shows, although life expectancy at birth was only 40 years

Table 9.1
Indicators of Levels and Changes in Health and Nutrition

		1960	1965	1970	1980	1985	1990	1992
Life Expectancy at birth	Male	–	41	43	–	49 (1988)	53*	50
	Female	–	43	46	–	52 (1988)	56*	53
	Both	40	–	–	49*	51	53*	52
Mortality rates per 1,000 live births	Infants (age< 1)	165	157	142	123*	111*	98*	99
	Child (1-4)	289	–	–	200*	162*	144*	179
Pop. per	Physician	–	33,200	31,720	–	23,850 (1984)	19,690	–
	Nurse	–	5,420	3160	–	2,640 (1984)	–	–
Daily caloric supply per capita	Total	–	2,092	–	2,107 *	2,225 (1988)	2,170 (1988-1990)	–
	% of requirement	–	92	–	84	89 (1998)	92 (1988-1990)	–
Access to safe water (% of pop.)		–	–	20 (1975-1980)	47 (1985 - 1987)	46 (1988-1990)	46	–
% of govt. spending on health		0.7	–	–	4.7*	4.9*	3.1 (1988-1990)	–
Adult illiteracy rates (%)	Male	–	–	–	–	52	50	37
	Female	–	–	–	–	65	62	60
	Both	–	–	–	–	74	55	49

Sources: UNDP, *Human Development Report 1994*; World Bank, *World Development Report 1991*; UNDP and World Bank, *African Development Indicators*.
*These figures are for the whole of Africa while the rest denote Sub-Saharan Africa.

in 1960 across the continent, by 1990 it had reached 53 years. Table 9.1 also shows a steady increase in life expectancy for both males and females, from 1960 to 1992. Likewise, Table 9.2 illustrates the significant African gains in life expectancy for seven periods, from 1950–55 to 1992. Life expectancy at birth

Table 9.2
Life Expectancy at Birth for Africa in Years

1950–1955	1955–1960	1960–1965	1965–1970	1970–1975	1975–1980	1992[a]
37.5	39.8	42.2	44.5	46.5	49.8	52

Note: [a]1992 is for sub-Saharan Africa.

Sources: [a]World Bank, *World Development Report 1993*, p. 163. The rest of the data came from Moone (1985, p. 9).

improved from 37.50 years in 1950–1955 to 44.5 in 1965–70 and then to 49.9 in 1980–1985.

Although life expectancy at birth for sub-Saharan Africa has progressively increased since the 1950s—when it was about 40 years—by 1960 it was still below 45 years depending on the source of data. Since independence, however, African countries have registered tremendous gains. By 1970, life expectancy had exceeded 45 years and by 1983 it exceeded 50 years. Thus, average life expectancy has increased by over one-third during the past three decades and overall access to health services has also increased. The World Bank projects sub-Saharan African life expectancy at birth to approach 60 years by 2020 alongside a significant decline in fertility rate from about 6.5 in 1990 to about 4 in 2020. But if the aggregate performance is reassuring, how have the individual countries fared?

From Figure 9.1, it is clear that all 48 countries recorded improvements in life expectancy during the course of the three-year period. By 1989, 60 percent of 48 African countries had achieved life expectancy of 50 years or more. Seychelles had a life expectancy of 70 years, the highest in 1989, while Guinea-Bissau had the lowest at 40 years. Remarkable improvements were achieved by Botswana, Egypt, Libya, Morocco, Swaziland, Tunisia, and Zimbabwe. Whereas there are 12 countries with life expectancy of 40 years or less in 1969, by 1979, there were only three. By 1989, all countries except Guinea-Bissau had exceeded 40 years. Ten countries had a life expectancy of 60 years or more in 1989, while 16 had between 50 and 60 years.

Infant Mortality

As with life expectancy, significant progress has been made in child survival in Africa over the past decades. The continent's infant mortality has declined from 165 per 1,000 live births in 1960 and 157 in 1965 to 98 in 1990. In other words, the survival of infants almost doubled in the post-colonial era. There is also a 50 percent decline in the infant and under-five mortality rates during the past three decades. From an average mortality rate of 191 per 1,000 population in early 1950s, the rate fell to 101 per 1,000 in the late 1980s. Perhaps, African countries have made the most remarkable health achievements in the reduction of infant mortality rates.

Figure 9.1
Life Expectancy in African Countries 1969, 1979, and 1989

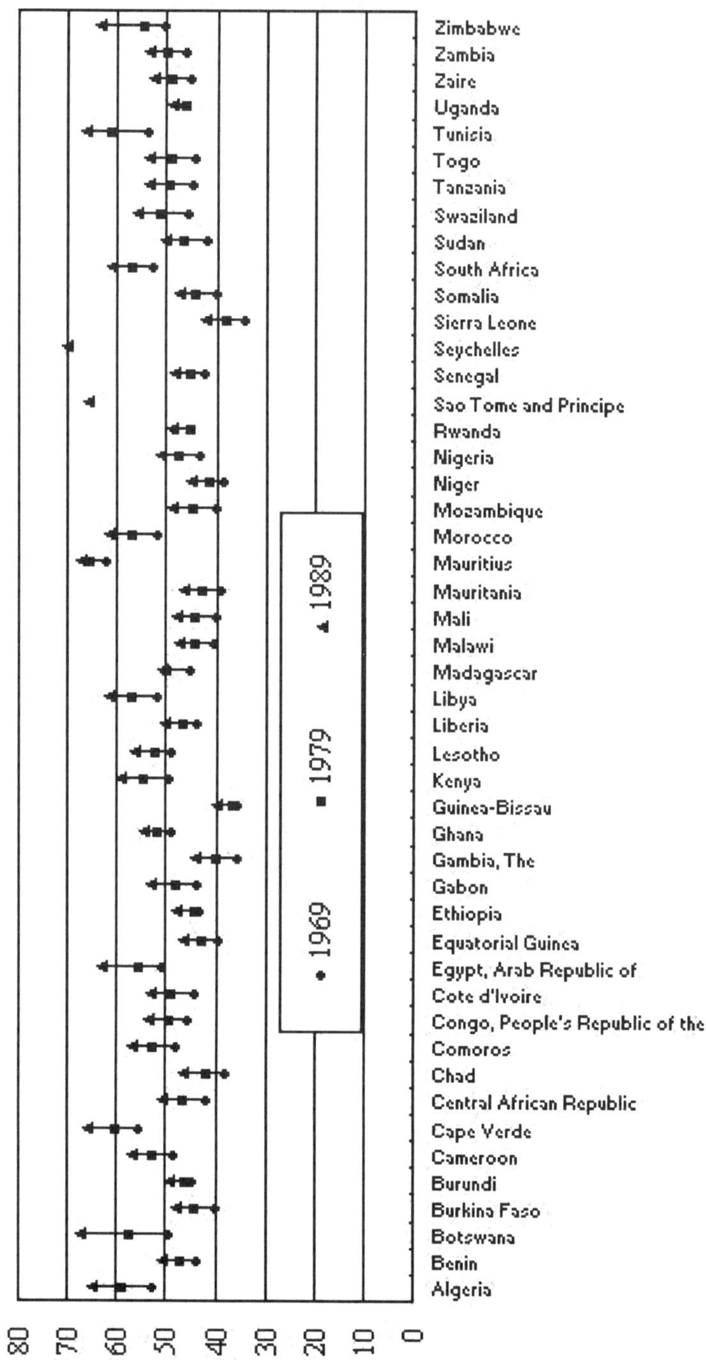

Figure 9.2
Infant Mortality Rates (per thousand) for 48 African Countries 1969, 1979, and 1989

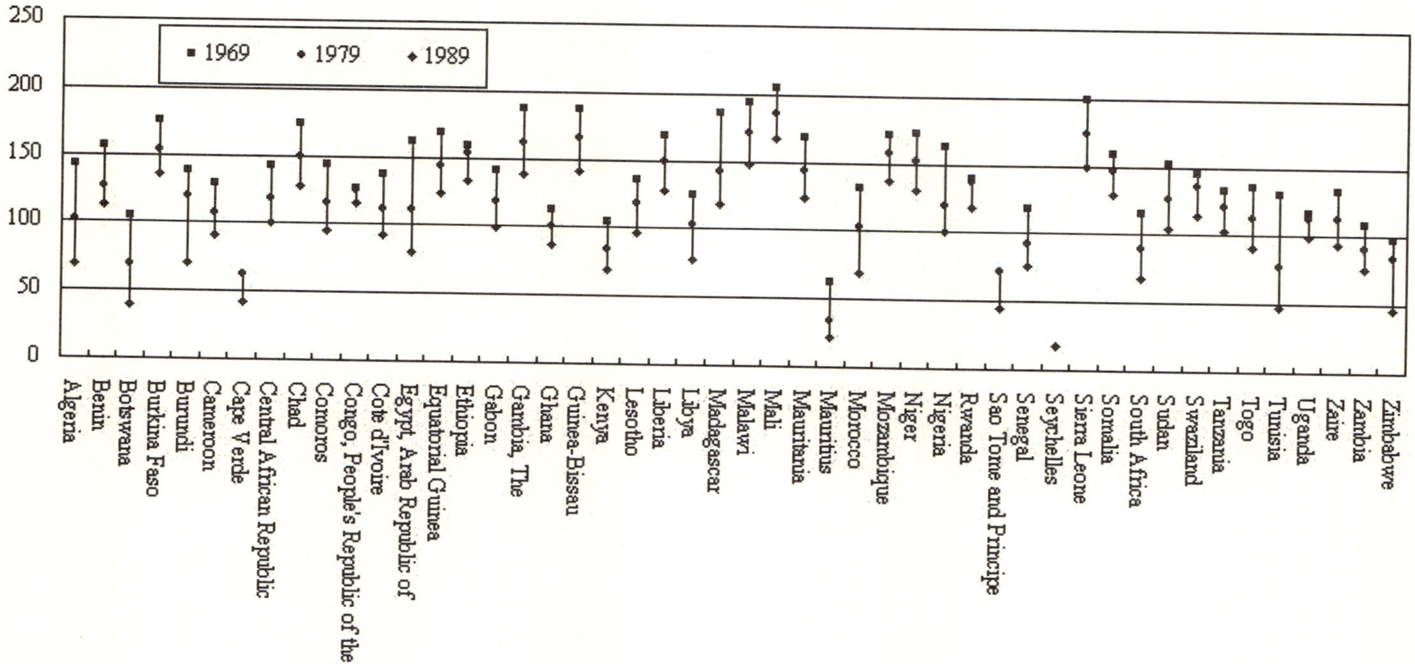

The comparison of 48 countries in Figure 9.2 show that by 1989, seven had attained an infant mortality rate of 50 or less per 1,000; 19 had 100 or less in 1989, compared to two in 1969. By 1979, all African countries had reduced their infant mortality rates to below 200. From 1960 to 1990, child mortality rates were reduced by more than 30 percent from 251 at the end of colonialism in 1960 to 171 in 1990. This decline in mortality rate and increase in life expectancy are accompanied by a decrease in fertility. Figure 9.2 also shows that in 1989, with the exception of Mali, all countries had improved their infant mortality rates to below 150 per 1,000 live births. The majority, 21 countries, fell within the 150–100 range. This means that more improvements are still needed to continue the reduction. Again, Seychelles takes the lead in life expectancy, followed by Mauritius. Notable improvements were recorded by Tunisia, Morocco, Madagascar, Egypt, Burundi, Botswana, and Algeria.

Population per Physician and Nursing Professional

Another area in which Africa has registered significant progress is supply of doctors and nurses. Between 1970 and 1990, the population per physician ratio in Africa fell from 31,730 to 23,540 (World Bank, 1993, p. 293). Though virtually all African countries have shown improvements, only those countries for which inter-temporal comparative data were available are shown in Figure 9.3.

As Figure 9.3 shows, Burkina Faso's numbers fell from 97,310 in 1970 to 57,310 in 1990, a 41 percent decline in the population to physician ratio in 20 years. This is, indeed, a significant accomplishment. Egypt and Madagascar have below 1,000. Ethiopia and the republics of Chad and Niger have also made major progress in the reduction of their population per physician ratio. With the exception of Senegal, Zambia, Egypt, Zimbabwe, Kenya, Madagascar, and Tanzania, all 20 countries for which data were available for the two reference periods 1970 and 1990, show significant reduction in their population per physician, which is due to an increase in the supply of doctors. Many of the countries that did not show changes already had low ratios. Between 1965 and 1980, population per physician declined 40 percent for Kenya, 62 percent for Sudan, 57 percent for Nigeria, 40 percent for Mali, and 35 percent for Senegal (Evlo, 1989, June).

Although data are sparse for population per nurse, available data show improvements for most countries (see, Figure 9.4). Tanzania had the most significant reduction in population per nurse, from more than 50,000 population per nurse in 1970 to below 500 in 1990. All the countries shown in the chart have below 1,000 population per nurse. This shows that more than one-third of African countries have greatly increased their supply of nurses relative to their population.

Daily per Capita Caloric Supply

Although daily caloric supply per capita has improved, it has also fluctuated between 1970 and 1990 (refer to Table 9.1). On the average, African countries

Figure 9.3
Population per Physician for Selected African Countries

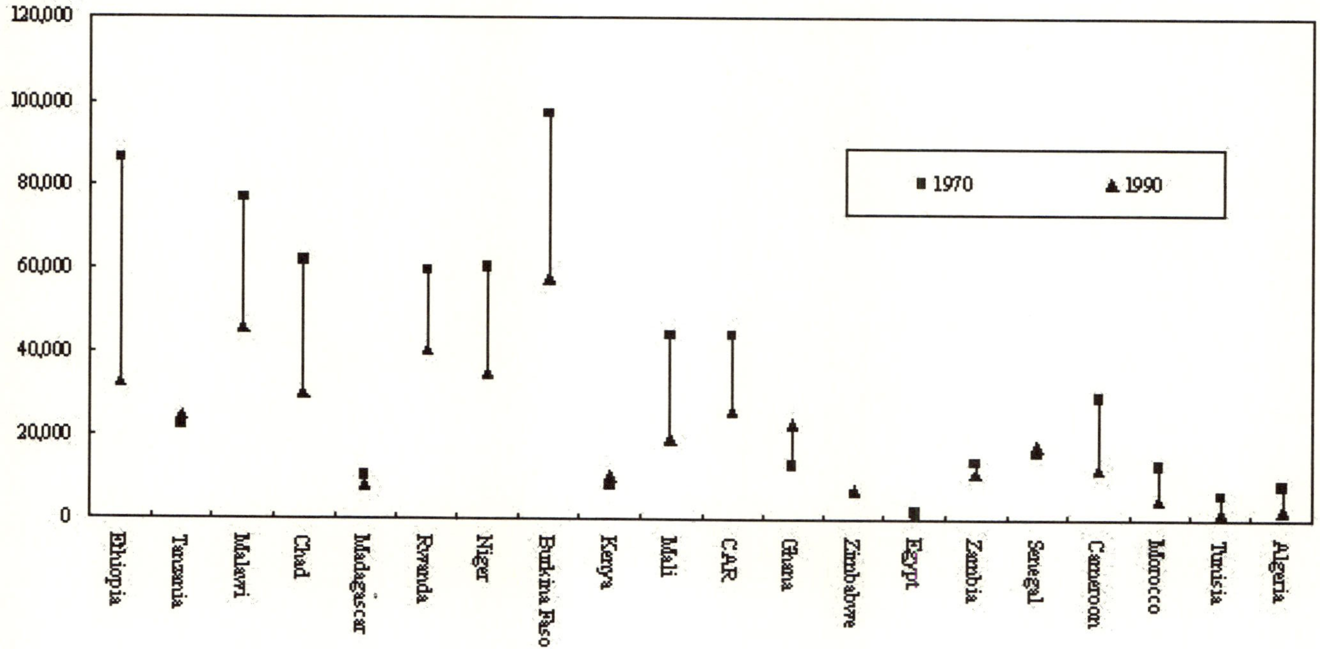

have supplied over 85 percent of the per capita daily caloric needs. Nutrition, especially child nutrition, has improved between 1965 and 1990. This improvement in child nutrition has been identified as one of the major contributions to the lowering of infant mortality rate. Daily caloric intake per capita is one of those variables that was greatly affected by the economic difficulties the continent suffered. It has, therefore, fluctuated over the years. Calorie supply per capita in 1988–90 for sub-Saharan Africa was 2,170 per day. As a percent of requirements, the daily calorie supply was 92 percent during the same period (United Nations Development Program, 1994).

Percentage of Population With Access to Safe Water

Globally, the percentage of families with access to safe water increased from less than 10 percent to almost 60 percent during the past two decades. As an economic infrastructural service, piped water helps to promote good health, thereby raising productivity and lowering production costs. World Bank data show that water supply expanded more in low-income countries (to which belong a majority of African countries) than in middle- or high-income groups. Actually, the share of the population in low-income countries with access to water reached 30 percent between 1975–1990. Table 9.3 shows rural and urban access to services in Africa for three periods 1985–1988, 1987–1990, and 1988–1990. Table 9.3 shows that for the three services of health, water, and sanitation, no appreciable improvements have been made over the course of these three periods. African rural access to water remained below 30 percent during the three periods. Nonetheless, as Table 9.3 also shows, significant rural-urban gaps still exist, but the results are mixed. In Ghana, only 1.5 percent of the poorest quintile had access to public water supply in 1987–1988 as opposed to 30.6 percent of the richest quintile. In Côte d'Ivoire, only 2.4 percent of the poorest quintile, as opposed to 62 percent of the richest quintile, had access to public water in 1985 (World Bank, 1994, pp. 4–5, 14–32).

Percentage of Central Government Expenditure on Health

Because of difficulties in tracking down what actually constitutes expenditure on health, data on African countries' expenditure on health are hard to come by, and when it is seen, the reliability is often questionable. Nevertheless, an examination of 11 francophone countries listed in Table 9.4 shows that their Ministries of Health's total annual budget increased 7.8 percent in 1971; 5.1 percent in 1972; and 10.5 percent in 1973. The picture is less rosy for the 1980s for almost all African countries.

On the average, expenditure on health in 1980–85 was 5.1 percent of total and 4.6 from 1986 to most recent. About 21 percent of the totals were spent on economic services (United Nations Development Program and World Bank, 1992). For the period 1980 to 1987, the total government expenditure as a ratio of total expenditure exceeded 4 percent for each of those years (Table 9.5). Figure 9.5 shows the total health expenditure as a percent of GNP in 1960 and 1970. Aside from Sudan, Uganda, and Nigeria, which recorded declines, all the

Figure 9.4
Population per Nurse for Selected African Countries (1970 and 1990)

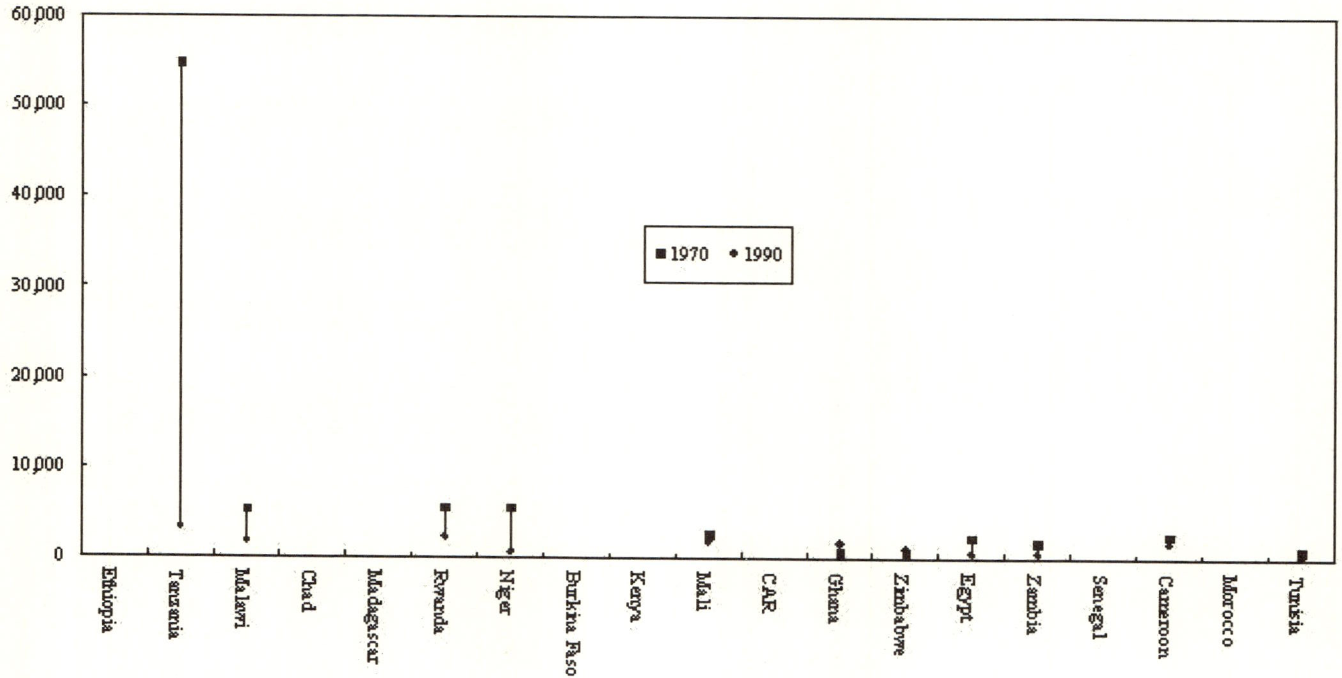

Table 9.3
Sub-Saharan Africa Rural and Urban Access to Services

	Services	1985–1988	1987–1990	1988–1990
Rural population with access to services (%)	Health	37	36	–
	Water	26	28	28
	Sanitation	18	17	18
	Percent rural population	69	69	69
Urban Population with access to services (%)	Health	72	80	87
	Water	74	65	79
	Sanitation	60	56	47
	Percent urban population	31	31	31

Source: United Nations Development Program, *Human Development Report 1994.*

Table 9.4
Annual Budgets of the Ministry of Health in Selected Francophone African Countries (in millions of Francs, CFA) 1970–1973

Country	1970 Amount	1971 Amount	Growth Rate (%)	1972 Amount	Growth Rate (%)	1973 Amount	Growth Rate (%)
Benin	1162.3	1274.5	9.7	1336.8	4.9	1424.1	6.5
Cameroon	2749	3008.9	9.5	3379	12.3	3777.1	11.8
C. African Republic	965.4	928.1	-3.9	1118.4	20.5	1161.1	3.8
Congo	1328.8	1547.1	16.4	1834.5	18.6	2396.3	30.6
Côte d'Ivoire	5808.4	6194.9	6.6	6318.6	2.0	7363.2	16.5
Gabon	1147.2	1487.9	29.7	1564.8	5.2	1730.8	10.6
Burk. Faso	853.9	894.5	4.7	864	-3.4	899.6	4.1
Madagascar[a]	2062.8	2242.2	8.7	2368.4	5.6	2899.1	22.4
Mali[b]	2038.2	1986	-2.6	2153.4	8.4	2246.7	4.3
Mauritania	480.4	509.6	6.1	541.6	6.3	648.1	19.7
Niger	874.6	898.8	2.8			1017.6	
Senegal[c]	3240.5	3409.5	5.2	3463	1.6	3446.5	-0.5
Chad	836.7	875.5	4.6	634.1	-27.6	680.1	7.3
Togo	573.2	684.3	19.4	797.6	16.6	880	10.3
Average Growth Rate			7.8		5.1		10.5

Note: [a]Madagascan Francs; [b]Malian Francs. [c]Senegal is for 1970/71, 1971/72, 1972/73, 1973/74.
Sources: *Le Financement du Système Sanitaire dans 14 Etats Africains et Malgache 1970–1974*, Institut d'étude du développement économique et social, unversité de Paris 1, 1976, p. 21.

countries shown increased their health expenditure. Botswana shows the highest increase followed by Libya, Cameroon, and Mauritania.

Table 9.5
Government Expenditure on Health as Percentage of Total

1980	1981	1982	1983	1984	1985	1986	1987
4.6	4.7	4.8	5	4.8	4.7	4.5	4.6

Source: UNDP and World Bank (1992). *African Development Indicators.* Washington, DC: The World Bank.

Evlo's (1989, June) data on seven African countries show that, with the exception of Zaire, all experienced significant declines in the rate of growth of real per capita government expenditure beginning 1981. This confirms the impact of the economic crisis of the 1980s on health expenditure. On the average 7.7 percent of the totals government expenditure went to health in Kenya between 1978 and 1983. Between 1977 and 1983, Malawi allocated 6.1 percent of total government expenditure to health; Mali spent an average of 4.6 percent; Niger, 4.5 percent; and Senegal, 3.8 percent. Much of the expenditure went to curative rather than preventive care (Evlo, 1989, June).

Accomplishments in Disease Prevention and Control

According to WHO, the major achievements in disease prevention and control relate to communicable diseases. "Thus the incidence of neonatal tetanus, poliomyelitis and measles has diminished through the medium of the expanded program on immunization" (World Health Organization, 1987, p. 27). Systematic immunization programs have reached more than two-thirds of African children and "Immunization coverage increased from less than 20% in 1985 to 57% in 1992 for DPT3." DPT3 are diphtheria, pertussis, and tetanus vaccines combined so that they can be administered to the child in a single injection. Also, nine countries reported "Zero incidence" of polio and others reported significant reduction in the disease (ACSI-CCCD, 1993, p. 11). Aside from achievements in immunization, diarrheal diseases and malaria have been significantly controlled. Child nutrition has also been improved. Reaffirming the overall achievements in disease prevention, Azevedo (1988, p. 119) writes that:

> Indeed, for the past 25 years, health facilities in Africa have been expanding at en ever-accelerating pace. African governments have attempted to provide free or low cost services to the sick and the poor. Furthermore, the effect of most debilitating diseases such as malaria, yellow fever, sleeping sickness, bilharziasis, cerebro-spinal meningitis, and others, has been reduced considerably. Some diseases such as smallpox have been eliminated altogether from the continent.

Consequently, in spite of serious world economic crises, wars, famines, and other social disasters during the 30-year period 1960–1990, the remarkable fact

Figure 9.5
Total Health Expenditure as Percent of GNP (1960 and 1987)

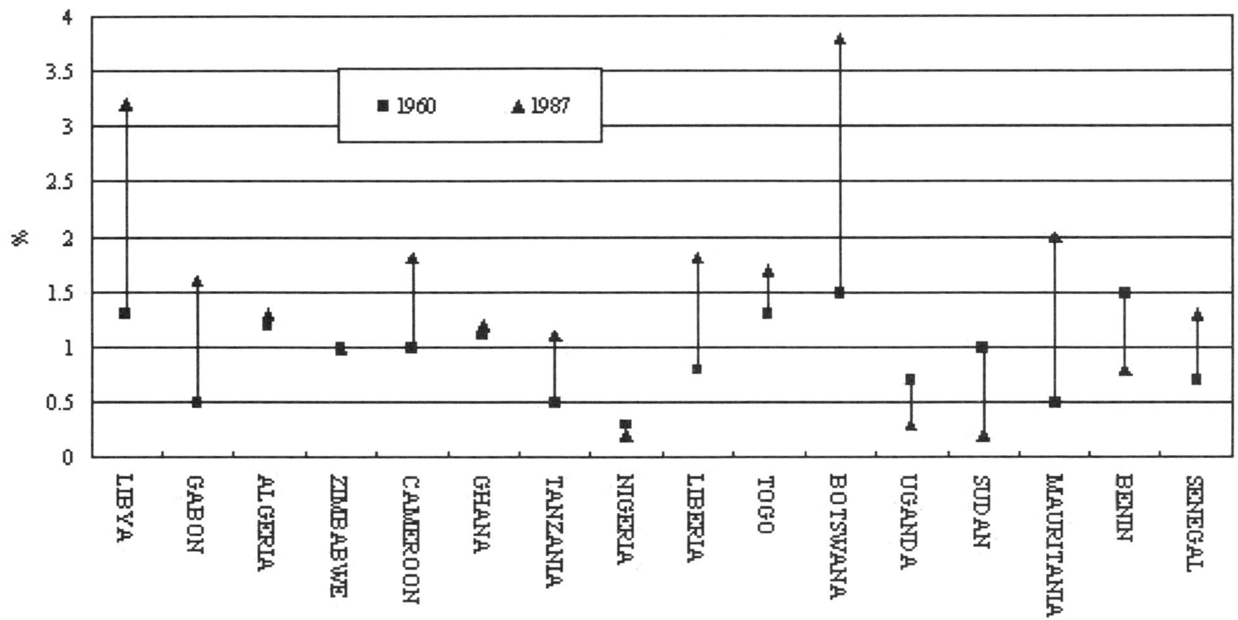

is that health status—insofar as it is reflected by life expectancy at birth—improved in Africa. Birthrates also declined from 47.9 in 1960 to 45.6 in 1980–1984. However, noncommunicable diseases such as cardio-vascular problems due to the demographic transition are on the increase. Africa may lag behind other regions, but improvements in post-colonial health conditions have been substantial. The significance of this historic performance is captured in the following statement in Roemer (1991, p. 27):

> These positive trends in the health of populations of virtually all countries should indicate that the countries' social efforts applied since World War II have not been in vain. Political rhetoric may stress the differences between rich nations and poor—and these differences, of course, are real—but the many strategies of governments to increase the accessibility of health services have had measurable effects. The substantially better health records of more economically developed countries should not obscure the fact that even in the least developed countries, improvements have been achieved.

HEALTH SYSTEM-BASED PERFORMANCE

Mobilization and Development of Health Resources

As shown in the previous section, the number of physicians has significantly increased in all African countries since independence. The vital contribution of traditional healers is also evident everywhere. The number and services of allied personnel, pharmacists, chemists, and roadside drug sellers has shown tremendous growth. Likewise, facilities have grown since independence in terms of more hospitals, clinics, and long term care. Growth in the supply of drugs, biologicals, medical supplies, and medical equipment is phenomenal. Furthermore, there is a widespread acceptance and use of public health interventions to help prevent and combat main causes of infant mortality. Consequently, there is widespread use of antigens to protect children against diseases, proper case management, growth in monitoring, breast-feeding, family planning, and prenatal care. The increase in the availability of health technology, telecommunications, sanitation, and water supply have had a positive net effect on the health sector. The reduction in population per physician ratio increased life expectancy at birth, and declining infant mortality rates are all indicators of improved resource mobilization and use in the continent.

Organizational Accomplishments

Better organization of health programs by African countries is evidenced by the establishment of ministries and departments of health where they did not exist at independence. Personnel training has accelerated all across the continent partly to compensate for the shortages that existed at independence. African governments have also established coherent health policies, by adhering to the international organizations such as WHO, and by adopting the WHO-established policies such as the Health-For-All by the Year 2000 program. Almost all countries have informal and less structured forms of social security programs and occupational health and safety programs thanks to increased labor unionization.

There is a greater recognition of the interrelatedness and connections between different branches of government that play health functions.

In recognition of this organizational achievements, Louis Sullivan, then U.S. Secretary of Health and Human Services, and Ronald Roskens state in their 1991 report to the President of the United States: "Consistent with the philosophy of the international public health community, African nations have increasingly focused on developing primary healthcare systems. As a result, in the last decade, remarkable progress has been achieved in increasing access to oral rehydration therapy, expanding immunization coverage and providing family planning services" (Sullivan and Rosken, 1991). The report acknowledges real progress in child survival noting that: "[H]ospital admissions due to diarrhea have been dramatically reduced due to management of diarrhea and dehydration with oral rehydration therapy. In several countries, measles no longer topped the list of causes of hospitalization and death because of successful immunization programs."

African leaders have also shown their interest in improving the health conditions of their citizenry by their participation in various summits. The commitment to make health accessible to all is expressed in the numerous resolutions of the WHO regional committee and of the Organization of African Unity. For instance, 1987 was declared the African immunization year to foster healthcare system development, effective utilization of the cold chain, and increased social mobilization and multisectoral collaboration in health (Hill, 1993). Following (Table 9.6) is a list of some of the activities that show commitment to the improvement of health on the continent.

Table 9.6
Activities Related to Health Improvement

Activity	Place	Year
Three-Phase Health Development Scenario	Lusaka	1985
Safe Motherhood Initiative	Nairobi	1987
Declaration by Heads of State and Government of OAU on Health as a Foundation for Development	Addis Ababa	1987
Special Health Fund for Africa	Brazzaville	1988
Bamako Initiative on Community self-financing of primary healthcare through provision of essential drugs and recovery of their cost	Bamako	1989
A new resolution of the Council of Ministers of OAU urging both urban and rural communities to set up mutual assistance or solidarity funds for health, to be eligible for support from special fund	Addis Ababa	1990

Source: World Health Organization, *Africa's Health: International Cooperation.* Geneva: WHO, 1991.

The UN has sponsored a World Summit for Children and the Rio Earth Summit in which African leaders participated. To indicate the strong commitment of African policy makers to desirable goals, various resolutions have been adopted over the years including those calling for the acceleration of expanded program on immunization, proclamation of 1986 as African immunization year, global eradication of poliomyelitis, and the elimination of neonatal tetanus; others include women's health and safe motherhood, and health of the newborn. As noted above, the OAU summit endorsed the "declaration of health as a foundation for development" at its 23rd ordinary session in 1993, as well as reaffirming full commitment to the goals of the 1990 World Summit for Children (Monekosso, 1993).

Earlier in 1978, at Alma-Ata, WHO proposed a coherent program aimed at Primary Healthcare (PHC). A good number of African countries subscribed to the PHC program. The advantages came in the form of refocusing the communities on their basic health units. The concept and application of PHC also increased the emphasis on health and a realization of the precarious health conditions by the authorities in the nations. PHC enabled communities to increase their participation and adopt their own health needs. From 1980, health services became re-energized by community participation in many African countries (Santé en Afrique, 1987, November).

On a different note, the enormous increase in the activities of voluntary bodies and the remarkable increase in the contribution of the private sector to healthcare are testimonies of African countries' efforts to improve healthcare services. In many countries, a greater ratio of moderate-sized to large enterprises are providing some form of healthcare either at the workplace or by special arrangement. This healthcare is usually provided either by someone trained to give first aid and in some cases arrangements are made with a nearby health facility to get physician services. In both rural and urban areas, traditional medical practitioners, physicians and dentists, pharmacists and drug store owners, and private hospitals are sprouting and providing services. In this writer's hometown, Nsu, Imo State of Nigeria, in 1989 there was only one clinic run by an absentee physician who came twice a week. By April 1994, there were a total of eight clinics all of them run by qualified residents and fully-trained physicians.

Increases in Economic Support for Healthcare in Africa

Roemer identifies five main economic support sources in a health system: tax revenue, social security, voluntary insurance, charitable donations, lotteries, foreign aid, health projects, individuals, and families (Roemer, 1991). A knowledge of the mix of these economic support sources generally helps in estimating how much importance a country attaches to health. Public and private expenditure on health as a percentage of GDP has grown over the years in all African countries. On the average, the percentage of overall government expenditure on the health sector was on the increase across the continent until the 1980s when global economic crisis, the debt burden, and structural adjustment depleted available resources. Likewise, external support for health has risen tremendously in real per capita terms since the 1960s, as foreign aid increased. The increase in the mix of economic support sources—tax revenues, charitable donations, foreign aid, and insurance—is an indication of the increasing importance African countries attach to health.

Multilateral and bilateral foreign aid has come in form of assistance given by technical personnel and advisors; equipment donation such as vehicles, medical and surgical equipment, vaccines, and medicine; training; and investment subsidies. UNICEF, which concentrates on children, has the highest proportion. In spite of this growing foreign aid, the distortions in terms of high-status versus low-status, urban versus rural are large. As the UNDP (1994, pp. 73-74) notes, "Urban hospitals get preference over primary healthcare, which gets less than 30 percent of total aid expenditure on health." The major sources of multilateral aid are WHO, UNICEF, *Fonds d'Aide et de Coopération* (FAC), USAID, IBRD/IDA, International Fund for Agricultural Development (IFAD), IDB, FAO, European Development Fund (EDF), African Development Bank, and the United Nations through its population activities.

On health insurance, Vogel (1990) shows that although the majority of African countries do not have formal health insurance systems, some countries have established coverages for their citizens. Out of 23 countries the author studied, 7 had formal health insurance with the percentage of population covered ranging from 0.001 to 11.4. Although the majority of the insured are government employees, there was also private insurance in 6 countries—Côte d'Ivoire, Ethiopia, Kenya, Nigeria, Swaziland, and Zimbabwe.

A major, but often ignored achievement, is the effectiveness of African governments in facilitating the work of nongovernmental health organizations (NGOs) and private voluntary organizations (PVOs) in their countries. DeJong (1991, p. 13) has noted that, "The work of the international NGO Save the Children, for example, often builds on existing government facilities and reinforces them in turn." DeJong states that in Mali and Ghana, for example, "governments frequently provide subsidies or personnel which enable missions to carry out their activities effectively and without which many would not survive."

Improvements in Health Management

The accomplishment recorded in mobilization and development of resources, organization, and economic support will not have been possible without significant achievements of African countries in planning, administration, regulation, and legislation. Virtually all countries present development plans or budgets filled with lofty health improvement goals for its citizenry. Though these goals are not usually fulfilled, they still reflect recognition of the important role health plays in social and economic development. There is also a recognition that achieving desirable health results depends on careful planning and articulation of health goals, effective administration of the healthcare sector, and the establishment of the regulations and legislation necessary to implement those goals. All the achievements recorded in the areas of declining infant mortality, growing life expectancy, generalized immunization are all manifestations of better health management by African countries. At issue here is the role of the state versus the role of the market. Marx, Durkheim, Polanyi, Weber, and Gerschenkron have all presented different views on the roles of the state. At one extreme is rent-seeking and predation typified by the absolutism of Mobutu Sese

Seko's Zaire and predicted by neo-utilitarian visions. At the other is the model of the activist developmental state (Evans, 1992). Therefore, the role of the state to accomplish desirable healthcare goals depends on its ability to overcome the constraints posed by the multiplicity of organizations and lack of capacity to pursue collective goals.

Enhancement of Health Delivery

Unlike colonial administrations, which emphasized the provision of curative services, African governments have, since independence, made significant strides in preventive services especially at the primary health level. This is evidenced by their adherence to WHO's Health for All by the Year 2000 program, which focused on primary healthcare (PHC). PHC addresses community health problems through "promotive, preventive, curative, and rehabilitative services" through the following eight actions: nutrition and education and adequate food supply, safe water and basic sanitation, maternal and child healthcare including family planning, immunizations, endemic disease prevention and control, treatment of injuries and diseases, and essential drug supplies (WHO, 1987).

As WHO's assessment shows, there have been significant improvements in health, education, promotion of good nutrition, improved sanitation, and the supply of safe water. Likewise, major achievements have been recorded in maternal and child care, including family planning, immunization against the major infectious diseases, prevention, and control of local maladies.

The success of PHC projects in Cameroon illustrates this trend. As Aurenche (1988) described them, the projects are based on the "service Catholique de la Santé" overseen by the bishops since 1959. The 1976 evaluation showed major accomplishments in the areas of maternal and infant protection, vaccinations, prenatal actions, village pharmacies, nutrition, and provision of health and sanitation infrastructure. Some health schools for the training of teachers were also established together with the integration of health in hygiene studies. Success was also recorded in the fight against leprosy and tuberculoses.

The accomplishments are not limited to preventive services. There have also been some improvements in therapeutic services through the provision of appropriate treatment of common diseases and the provision of essential drugs. Secondary care through ambulatory medical services, hospitals, and non-medical specialists have developed and improved in some countries. In contrast, due to obvious financial limitations, tertiary care in the form of highly specialized interventions are only limited to a few countries such as Nigeria, South Africa, and Morocco. Coles (1993, p. 1) writes, for instance, that, "Access to clinic-based treatment of diarrhea with ORS was 57% in 1991, up from 22% in 1985." Africa's DPT3 coverage was 57 percent in 1991, demonstrating that more than half of Africa's infants are seen by healthcare workers at least three times during their first year of life.

To give a picture of how African countries have performed comparatively, we use the World Bank's criteria to classify African economies and broadly distinguish them on the basis of per capita income: low-income, $635 or less (34 economies); lower middle-income, $636 to $2,555 (11 economies); and upper middle-income, $2,556 to $7,910 (six economies) (World Bank, 1991). For the low-income group, out of 28 African countries listed by the World Bank's *World Development Report 1994,* 19 had declining crude birthrates per 1,000 population between 1970 and 1992; 7 actually had higher rates and two remained unchanged. Among the middle-income group, 7 out of 8 had lower birthrates between 1970 and 1992; only 1, Congo, recorded an increase. All 4 African countries belonging to the upper-middle income group for which data were available, had declining crude birthrates and declining crude death rates. Average crude birthrate in sub-Saharan Africa was 47 in 1970 and 44 in 1992. Crude death rate fell from 20 in 1970 to 15 in 1992. Total fertility rate has also declined over time: 6.5 in 1970 and 6.1 in 1992; it is expected to drop to 5.6 in 2000.[1]

The low-income group also registered significant reductions in death rates, since 27 out of 28 countries that belong to this category showed declines between 1970 and 1992. The same is true of the lower middle-income group where 6 out of 7 had declining crude death rates from 1970 to 1992. Only 1, Congo had a rate of 6 for both periods. For sub-Saharan Africa, crude death rate per 1,000 population dropped from 20 in 1970 to 15 in 1992. With regard to infant mortality rates per 1,000 live births, 22 low-income-countries have lower infant morality rates in 1992 compared to 1970, while 5 countries had increased infant mortality rate.

For the 52 countries listed in the *World Development Report 1994*, per capita health spending data were missing for 18 while five had very spotty information. Botswana, Lesotho, Mauritius, Swaziland, Ghana, Zimbabwe show progressive growth. Cameroon, the Gambia, Guinea Bissau, Madagascar, Malawi, Niger, Rwanda, Tanzania, Togo, Uganda, Egypt, Zaire, Zambia, and Tunisia show fluctuating growth. Ethiopia was unchanged at 1.1 and 1.4 percent. Those that recorded declines were Kenya, Liberia, Nigeria, Senegal, Sierra Leone, Somalia, and Morocco. For all income groups, the variables that should be declining are indeed declining; those that should be increasing are increasing. The exception is the maternal mortality rate which remained largely unchanged for the low-income African countries, indicating the need for policy to focus on this area.

The Global Strategy of Health for All by the Year 2000

Ten years after establishing the Global Strategy for Health for all by the Year 2000, second evaluation and 8 reports on the world health situation shows several accomplishments and problems (Table 9.7). WHO (1993, pp. A31 & B4) attributes the limited ability of member states to implement the health-for-all strategy more effectively to "weak management performance, inadequate training and utilization of human resources for health, failure to integrate health research and insufficient application of appropriate technologies to the health development process." No mention was made of the impact of structural adjustment. According to WHO's overall assessment in terms of traditional indicators—life expectancy and infant and child mortality—health status is improving globally. Developing countries "have reported a general improvement in nearly all the global indicators of health development and health status" (WHO, 1993, p. A31). Areas of achievement include an increase in the coverage for immunization against six vaccine-preventable diseases such as DPT and poliomyelitis vaccines. Consequently, preventable diseases among children are showing a noticeable decline. Availability of essential health services is on the increase. Health is increasingly being recognized as a basic element of development. Literacy rates are rising and women's role in development is being recognized.

Table 9.7
Accomplishment and Problems

Accomplishments	Problems
Infant mortality continues to decline.	Malaria and AIDS persist as public health problems.
Considerable progress has been made toward the organization of health systems in line with the primary healthcare approach.	Nutritional status especially for children remain precarious.
District health systems have become a reality in almost all countries.	Problems still exist with available local healthcare and essential drugs.
Immunization coverage has progressed steadily.	Key elements of child survival and safe motherhood programs are still not adequately integrated in national programs.
Spectacular improvements have been recorded in access to immunization services.	Budgetary restrictions have often resulted in a reduction in operating expenditure and investment.
Access to drinking-water especially in urban areas has greatly improved.	Imbalances and inequality in the provision of health services have deteriorated in some countries.
Advancement in the area of knowledge and health technology have been registered.	Formation of policies on human resources remain the exception rather than the rule.

THE ROLE OF THE TRADITIONAL HEALERS

Various colonial administrations declared traditional medicine illegal—underscored by Keen's (1960, pp. 15–16) claim that traditional medicine is "an evil system that can and must be overcome" and over which the "principles of Western medicine will ultimately triumph." In Niger, for instance, during the almost 60 years of French colonization, the colonial administration declared war against traditional healers and spiritual priests, eliminating them with military force. Following colonial precedence, such countries as Kenya and Côte d'Ivoire declared traditional medicine illegal after independence. Notwithstanding these historical antagonisms, the reality of Africa is that traditional medicine still plays an important role as it has done over the centuries. Traditional healers treat the patients who are not treated by the modern health sector. They are available and accessible in remote areas; they are also affordable and their mobility allows them to operate within the realities of the population they serve. These faith healers, herbalists, and spiritualists are successful because, in addition to healing, they help in disease prevention, establish personal contact, treat communicable and psychosomatic diseases "which demand thorough understanding of personal and sociocultural factors in the individual contact with the patient" (Staugård, 1985, p. 178).

In the case of Northern Nigeria, Stock (1980, p. 432) shows that, "Traditional healers are settled and respected members of the community, and have both a proven commitment to the practice of healing and practical experience in dealing with the sick." Likewise, Staugård's study in Botswana is illuminating as it highlights the considerable role the traditional healer plays in an African society. This study illustrates several interesting points. First, it shows that a high proportion of visits is to receive traditional care for preventive reasons. Second, it shows that the traditional system seems to be more preoccupied with prevention of ill health and disease than the modern sector. Third, the study indicates that traditional healers were very valuable in dealing with psychosomatic complaints and communicable diseases. Finally, according to Staugård's study, 68.3 percent of the rural population and 30.9 percent of the urban population studied had visited a traditional healer on one or more occasions in a 12-month period (Staugård, 1985, p. 178).

Both Stock (1980) and Good (1988) have stressed the positive role of the traditional healer. On the one hand, Stock (1980, p. 422) demonstrates through his study of healthcare behavior in rural northern Nigeria that, "The per capita utilization of all types of traditional therapies increased significantly with distance from healthcare facility." He adds that traditional healers were consulted for the treatment of serious and persistent sickness. On the other hand, Good highlights the importance and overwhelming role of traditional healers (THs) in provision of African medicine. Citing Anyiam (1987) and Nchinda (1976), Good (1988, pp. 65–67, 68) writes that: "Generally speaking, THs will continue to be a popular option because they are accessible, available, acceptable, and adaptable. They survive because they represent a culturally understood approach to managing physical and psychological ills."

ACCOUNTING FOR THE SUCCESS IN HEALTH

Several factors are responsible for the African governments' achievements in health: growing incomes, improved education, government's efforts to expand health services, and enrichment of these efforts by technology (World Bank, 1993, p.1). Most of this gain in health status is accounted for by control of communicable diseases especially during childhood. More specifically, the World Bank attributes the dramatic and unprecedented decline in mortality rates to three factors: income growth, which meant more food for the poor, improved housing, and more healthcare; improvements in medical technology; and public health and the spread of knowledge about clean water, sanitation, and food. For instance, the decrease in Senegal's infant mortality rate in the past 10 years from 116 per 1,000 to 84 per 1,000 is "due in part to improvements of basic healthcare services, to the satisfactory vaccine coverage rate (80% and more by antigen), to the combined fight against diarrheal diseases and respiratory infections, and to a great number of programs" (ACSI-CCCD Report, 1993, p. 6). Most of the gains in African child survival have resulted from improvements in survival after the neonatal period.

PROBLEMS

The decade of the 1980s, Africa's lost decade, was marked by significant declines in health budgets because of economic crisis and the adoption of economic reform programs. With major cuts in social expenditure, there was a decline in public expenditure in health in the poorest countries by 60 to 70 percent. This caused the percentage of GNP spent on health to drop below what it was in 1960s (ACSI-CCCD Report, 1993, p. 10). The main problems resulting from this funding inadequacy are high absolute levels of mortality exemplified by high levels of child mortality rates and high maternal mortality rates. Another problem is the increased disease burden measured in terms of loss of productive years or "the disability adjusted life year" (DALY) (World Bank, 1993). Such losses have been intensified by the incidence of HIV and AIDS and the increasing problems of drug-resistant malaria parasites, noncommunicable diseases due to increased aging of the population, and substance abuse, especially tobacco.

There are also problems with the existing health systems in terms of insufficient public healthcare services and clinical care, misallocation toward low cost-effective areas, and inequality in government allocation of health expenditure in terms of rich and poor (World Bank, 1993). Associated with these problems is the inefficiency and wasteful spending on drugs and underutilization of health workers, and hospital beds an exploding cost of healthcare provision. According to an analysis by WHO, there are problems with the development and management of human resources, lack of information, absence of clear policies toward equitable resource allocation and utilization. Other problems relate to the slowness in changing education to focus on prevention and health promotion, lack of interest and commitment of the health profession, lack of emphasis on research and technology uses, and absence of infrastructure (WHO, 1993).

Economic reforms and adjustment imposed by the international financial institutions (IFIs) have affected gains in healthcare as they have more adversely affected the poor because of unequal sharing of adjustment-induced pain and gain (Stallings, 1992). Amid acute declines in real wages, for instance, the rich are the least affected given that they have more options to reduce risks. Unfortunately, the impact is immense on the young children, pregnant women, and nursing mothers—the most vulnerable members of the society. These realities, along with the scourge of AIDS, have set back post-colonial African achievement in healthcare.

CONCLUSIONS

In spite of the historical colonial neglect of the healthcare sector and the economic constraints Africa has faced, achievements in post-colonial Africa have been substantial in the area of healthcare. The health status has improved in terms of appreciable increases in life expectancy, significant decline in infant mortality rates, reduction in the population per physician and nurse ratios, and improved nutrition. Likewise, resource mobilization, organization, economic support, management, and health delivery have also improved. Precisely, post-colonial African performance can be summarized as follows: significant extension of life expectancies; significant reduction in infant mortality rates; notable increases in the supply of doctors and nurses; lackluster performance in the improvement of nutrition; mixed performance in health spending, and poor performance in rural access to water.

Colonialism kept healthcare systems in a rudimentary state, which various African indigenous governments decolonized and improved on. In spite of the notable post-colonial improvements, emphasis is often laid on the problems of Africa's healthcare system, and the role of traditional medicine is almost completely ignored. Yet traditional healers, who sometimes outnumber physicians, have contributed significantly to Africans' health. The challenge facing African governments in the twenty-first century is to maintain the post-colonial momentum despite the problems of economic adjustment and the ravages of AIDS.

NOTE

1. The World Bank has adopted the use of prevalence of malnutrition in children under 5, believed to be a better measure of nutritional status than the previously used indicator, daily calorie supply. Because this chapter is analyzing historical and time-series data, we have used the daily calorie supply. Since the analysis do not mainly compare Africa with other regions, this decision would not adversely affect conclusions. The countries listed here may not represent politically independent territories but represent countries with distinct social and economic statistics (See World Bank, 1994).

REFERENCES

Africa Child Survival Intelligence (ACSI) and Combating Childhood Communicable Diseases (CCCD) Project. (1993). *Africa's progress in child survival.* Report prepared by ACSI-CCCD based on a forum, Dakar, Senegal, March 29–April 2, 1993. Washington, DC: ACSI-CCCD.

Anyiam, C. A. (1987). *Persistence with change: A rural urban study of ethnomedical practices in contemporary Ghana.* Unpublished doctoral dissertation, Department of Geography, Queens University.

Aurenche, C. (1988). Bilan et perspective en 1984 d'un projet de soins de Santé Primaire au Cameroon. In R. W. Carlaw & W. B. Ward (Eds.), *Primary health care:*

The African experience (pp. 287–314). Oakland, CA: Third Party Publishing Company.

Azevedo, M. (Ed.). (1988). *Cameroon and Chad in historical and contemporary perspectives* (African studies volume 10). Lewiston, NY: Edwin Mellen Press.

Behrman, J. R. & Deolalikar, A. B. (1988). Health and nutrition. In H. Chenery & T. N. Srinivas (Eds.), *Handbook of development economics vol. 1* (pp. 631–711). Amsterdam: Elsevier.

Chuke, P. O. (1988). Nigeria. In R. B. Saltman (Ed.), *The international handbook of health-care systems* (pp. 229-240). New York: Greenwood Press.

Coles, J. E. (1993). In *Africa's progress in child survival.* Report prepared by the Africa Child Survival Intelligence (ACSI) and Combating Childhood Communicable Diseases (CCCD) Project, based on a forum, Dakar, Senegal, March 29–April 2, 1993. Washington, DC: ACSI-CCCD.

Coquery-Vidrovitch, C. (1989). Les populations africaines du passé. In D. Tabutin (Ed.), *Population et societés en Afrique au sud du Sahara* (pp. 51–72). Paris: Harmattan

DeJong, J. (1991, June). *Nongovernmental organizations and health delivery in sub-Saharan Africa. Policy.* Washington: DC: The World Bank.

Evans, P. (1992). The state and the politics of adjustment. In S. Haggard & R. Kaufman (Eds.), *The politics of economic adjustment: International constraints, distributive conflicts, and the state* (pp. 145–151). Princeton, NJ: Princeton University Press.

Evlo, K. (1989, June). *Health finance in Africa: Tracking the aid experience of the child survival emphasis of countries.* Prepared by the Research for Child Health (REACH) Project.

Good, C. M. (1988). *The community in African primary health care: Strengthening participation and a proposed strategy.* Studies in African health and Medicine vol. 2. Lewiston, NY: The Edwin Mellen Press.

Hill, T. (1993). The health of the African child: Deliberations of the World Summit on the Child and the Year 2000 Goals. *Africa's progress in child survival.* Report prepared by the Africa Child Survival Intelligence (ACSI) and Combating Childhood Communicable Diseases (CCCD) Project, based on a forum, Dakar, Senegal, March 29–April 2, 1993. Washington DC: ACSI-CCCD.

Ityavyar, D. (1992). The colonial origins of health care services: The Nigerian example. In T. Falola & D. Ityavyar (Eds.), *The political economy of health in Africa* (pp. 65–87). Athens: Ohio University, Center for International Studies.

Keen, P. (1960). *Western medicine and the witchdoctor.* Johannesburg: The Institute for the Study of man in Africa.

Mburu, F. M. (1992). The impact of colonial rule on health development: The case of Kenya. In T. Falola & D. Ityavyar (Eds.), *The political economy of health in Africa* (pp. 88–106). Athens: Ohio University, Center for International Studies.

Monekosso, G. L. (1993). In *Africa's progress in child survival.* Prepared by the Africa Child Survival Intelligence and Combating Childhood Communicable Diseases Project, based on a forum, Dakar, Senegal, March 29–April 2, 1993. Washington, DC: ACSI-CCCD.

Moone, J. (1985). *The physical qualities of life in sub-Saharan Africa: A survey of a twenty-two nation fact finding mission* (1st ed.). Lawrenceville, VA: Brunswick Publishing Co.

Nchinda, T. C. (1976, July). Traditional and Western medicine in Africa: Collaboration or confrontation? *Tropical Doctor*, pp. 133–135.

Pearce, T. O. (1992). Health inequalities in Africa. In T. Falola & D. Ityavyar (Eds.), *The political economy of health in Africa* (pp. 184–216). Athens: Ohio University, Center for International Studies.

Roemer, M. I. (1991). *National health systems of the world,* vol. 1, The countries. New York: Oxford University Press.

Santé en Afrique: Perspectives et stratégies de coopération. (1987). Rapport du séminaire tenu à Lomé, République du Togo du 1 au 6 novembre.

Shapiro, M. F. (1983). *Medicine in the service of colonialism: Medical care in Portuguese Africa 1885–1974.* Unpublished doctoral dissertation, University of California, Los Angeles.

Stallings, B. (1992). International influence on economic policy: Debt, stabilization, and structural reform. In S. Haggard & R. Kaufman (Eds.), *The politics of economic adjustment: International constraints, distributive conflicts, and the state.* Princeton, NJ: Princeton University Press.

Staugård, F. (1985). *Traditional medicine in Botswana: Traditional healers.* Gaborone: Ipelegeng Publishers.

Stock, R. F. (1980). *Health care behavior in a rural Nigeria setting.* Unpublished doctoral dissertation, the University of Liverpool.

Sullivan, L. W. & Roskens, R. W. (1991). *Child survival and AIDs in sub-Saharan Africa: Findings and recommendations of the Presidential Mission to Africa, January 4–18, 1991.* Washington, DC: Department of Health and Human Services & U.S. Agency for International Development.

Twumasi, P. A. (1982). Improvement of health care in Ghana: Present perspectives. In P. S. Yoder (Ed.), *African health and healing systems: Proceedings of a symposium* (pp. 199–215). Holyoke, MA: Crossroads Press.

—— (1992). Colonial rule, international agency, and health: The experience of Ghana. In T. Falola & D. Ityavyar (Eds.), *The political economy of health in Africa* (pp. 107–120). Athens: Ohio University, Center for International Studies.

United Nations Development Program. (1994). *Human development report 1994.* New York: Oxford University Press.

United Nations Development Program and World Bank. (1992). *African development indicators.* Washington, DC: World Bank.

Vogel, R. J. (1990, August). *Health insurance in sub-Saharan Africa: A survey and analysis.* Washington, DC: World Bank.

World Bank. (1991). *World development report 1991.* New York: Oxford University Press.

—— (1993). *World development report 1993.* New York: Oxford University Press.

—— (1994). *World development report 1994.* New York: Oxford University Press.

World Health Organization. (1987). *Evaluation of the strategy for health for all by the year 2000.* Seventh Report on the World Health Situation, vol. 2, African region. Brazzaville: WHO Regional Office for Africa.

—— (1993). *World health status annual 1992.* Geneva: WHO.

10

Biotechnology, Food Production, and African Advancement

Richard W. Hull

This chapter is written to counter the corrosive disease of Afro-pessimism. It argues that Africa is not on the verge of collapse. In fact, exciting regenerative forces are at work, especially at the grassroots level, which hold enormous promise for Africans in the twenty first century. By viewing Africa through paradigms of positivism and by examining the micro-dimensions of sustainable development, Africa can present to itself and to the world a more balanced and optimistic profile. It is time to engage Africa more vigorously and dynamically, especially in the area of biotechnology.

This chapter argues that biotechnology and other innovations are already playing a crucial role in revolutionizing agricultural productivity and in preparing Africa to regain its self-sufficiency in food production. We cannot dismiss the warnings of Michael Barratt Brown and others that biotechnology poses a serious threat to the future of Africa's agricultural exports because of its potential, through genetic engineering, of producing cheap synthetic substitutes (Brown and Tiffen, 1992). Brown contends that biotechnology gives manufacturers much greater flexibility in their choice of primary commodity inputs. This may indeed be true, but biotechnology also holds enormous promise for African advancement. As world agriculture begins to rely less on chemical inputs, traditional African techniques of organic farming, utilizing genetically-engineered super-crops, may increase both the volume and quality of food.

The U.S. mass media tends to ignore the contributions of biotechnology to African advancement. The subject does not fit the framing of news for a public that is presumed to be interested primarily in news of conflicts and suffering. Indeed, analyses of African performance often overlook the direct relationship between science and socio-economic growth. Technology and science exist in a symbiotic relationship: technology utilizes the advances of science to produce newer and superior products; yet advances in technology depend increasingly on basic scientific knowledge (Zymelman, 1990, p. 3).

In less than a decade, scientists made enormous strides in our knowledge of basic biochemistry and plant physiology. In Africa and elsewhere they are making impressive contributions toward an understanding of the chemistry that integrates our living world into a synergistic interdependent whole. Biotechnology can be used to promote the intensification of sustainable agricultural-production systems and to improve environmental-management processes. Through genetic engineering, cells and molecules can be manipulated to alter the very building blocks of living organisms.

We are witnessing worldwide an extension of plant research from applied studies on problems facing specific crops to basic lab research on plants that may have no agricultural value but that can provide clues to gene manipulations that can be transferred to crops of vital importance to human and animal nutrition. Seemingly worthless but fast-germinating weeds can serve as mice in laboratory experiments.

Biotechnology tends to be extremely expensive, requiring sophisticated labs and highly-trained personnel. But its applications can often be simple and inexpensive and may bring surprisingly profitable results. For years, genetic engineering has been almost exclusively the preserve of multinational pharmaceutical corporations in the western hemisphere. But recently, biotechnology has begun to focus attention on the needs of the Third World, if for no other reason than to give the big pharmaceutical companies access to the rich genetic resources of the tropical rainforests. In the last decade, genetic engineering has opened up exciting new frontiers for food production south of the Sahara. Forgotten crops are being revitalized as little-known food plants and pre-colonial agricultural systems are being rediscovered. There is new respect for indigenous knowledge which could be a potent force in agricultural regeneration.

Traditional crops are being reconstructed to resist certain insects and viruses, and other crops are being modified to alter their nutritional properties, to increase their yields, or to make them more resistant to drought. Presently, genetically- engineered crops are being tested in many African environments and with surprisingly positive results. For example, through work in cell biology and molecular genetics, cells have been retrieved from the best plant varieties to create a "super race" of vegetables or hybrids.

The use of biotechnology for plant propagation and animal breeding is beginning to raise levels of crop productivity and overall food production. Advances in genetically-enhanced storage characteristics of perishable leguminous crops have enabled African produce to reach more distant markets, and the transferability of embryos has helped to breed animals resistant to such deadly diseases as trypanosomiasis. As a result, trypano-tolerant cattle have begun to penetrate new pasturage and farmers have expanded herds (Kesseba, 1989, p. 360). Mortality rates of cattle have also been drastically reduced through the use of the medicine Dectomax, introduced into southern Africa in 1993 by Pfizer, the international pharmaceutical giant. Dectomax has an extraordinarily broad spectrum of activity against parasites in subtropical cattle (Pfizer, 1993, p. 24).

HIGHER YIELDS FROM TRADITIONAL STAPLES

African-based research institutions are developing a wide range of new food varieties that could end the continent's food insecurity. New varieties of African staples developed in the laboratory are beginning to prove their worth in the fields.

Advances in plant technology have been especially impressive with cassava, which is arguably the greatest source of food energy in Africa. Cassava, an important food starch that is also used in the production of alcohol, grows chiefly in marginal soils and is a key crop for many resource-poor farmers. The crop has the advantage that it can be stored underground until it is needed for drying and processing into cassava flour, known as garri. The Ibadan-based International Institute of Tropical Agriculture (IITA) developed a new strain of super-cassava which offers the prospect of substantially higher yields. Indeed, IITA has produced six new strains of cassava that have doubled the yields of existing varieties and are more resistant to disease. African cassava production has begun to boom in many parts of the continent. Output in Nigeria alone rose by more than two-thirds between 1990 and 1993 (Bryant, 1993, pp. 49–50).

Africa is also experiencing a major cereal crop breakthrough in sorghum, a crop of great antiquity. A new hybrid, developed in the mid-1980s in the Republic of Sudan, yields up to 150 percent more than traditional varieties, can tolerate a broad range of soils, and may survive severe droughts where its predecessors could not (U.S. Department of State, 1985). Maize, an ubiquitous crop of American provenance, is also undergoing a transformation. In 1990, plant breeders at the International Maize and Wheat Improvement Center in Mexico created a new variety of corn that overcomes two major limitations of growing the crop in many parts of Africa: drought and acidic soils. The new breed can form tassels and silks at the same time under conditions of prolonged drought. Plant breeders have also come up with corn breeds with special roots that grow in highly acidic soils where corn has previously withered. Already in several African countries these new corn varieties are outyielding conventional breeds by upwards of 10 percent (*Washington Post*, 1994, p. A3).

New developments are also happening to the venerable palm tree. Some farmers rank palms as the world's third most useful plant family (after grasses and legumes), and for good reason. Palms provide edible fruit, oils for soap and machine lubrication, sap for beverages or sweeteners, leaves for thatching and basketry, wax for candles, and feed for livestock. The introduction of in-vitro tissue culture techniques for the oil palm in West Africa has boosted yields considerably (Kessaba, 1989, p. 361).

Impressive advances are also being made in grasses for livestock pasturage. A unique grass called vetiveria zizanioides prevents soil erosion and conserves water; it also vastly reduces rain runoff and could replace higher maintenance conservation systems (Bennetzen, 1993, pp. 259–260).

NEW CROPS IN OLD FIELDS

Africa has been a recent witness to the introduction of new crops as well as to new uses for traditional animals. In our century, camel's milk has been neglected in arid regions as a readily available source of fluid and protein. Some sahelian countries are now promoting programs to encourage extensive camel farming in dry areas. Unlike cattle, the camel can feed on thorny and saline vegetation that cattle find indigestible. It has been estimated that milk from a healthy camel can keep up to 20 children alive under conditions of severe prolonged drought (Wainaina, 1991, p. 35).

A winged bean, a new crop, was introduced to Africa in the early 1980s from southeast Asia. It is highly nutritious (especially high in vitamin A), easily grown, and nearly every portion of it is edible. Although some societies have rejected the bean for its coarseness and because it is seen as a food for the poor and primitive, it is gradually gaining wider acceptance (Schultes, 1981).

Even more spectacular has been the rapid progress made by molecular biologists in using genetic engineering to boost rice production. Biologists have transferred DNA from bacteria into rice protoplasts and come up with a prototype breed of super-rice that yields 25 percent more food per acre than the best varieties—and with considerably less fertilizer. West Africans have been cultivating rice for centuries and are reputed to have introduced it to the southern United States during the slave era. Their long experience with rice in West Africa's river valleys under harsh environmental conditions is now being put to good use with new species.

BIOPESTICIDES

Interest in biological control agents in fighting pests is the outcome of growing insect resistance to chemical pesticides and deepening concern over the environmental consequences of continued use of chemicals. According to P. J. Warner, biopesticides have considerable advantages over those derived synthetically. They are less expensive to produce and, more important perhaps, is their specificity. "[I]n most cases they are derived from a pathogen of a particular host (pest) . . . leaving other beneficial insects, birds, and mammals unscathed" (Warner, 1992, p. 94).

The advent of molecular biology has made possible the development of novel strains of microbes for use in agricultural processes. Innovative applications of microorganisms in the field have produced startling results. As we have seen, they include the creation of improved strains of food crops, but they have also produced entirely new strains of lactic acid for food and feed fermentation and for the maintenance of animal health.

Globally, farmers have become conscious of the expense as well as the health and environmental dangers of chemical pesticides and fertilizers. Africans are returning to the wisdom of their ancestors who understood that any natural ecosystem contains both pests and their natural predators. Scientists in Africa for the last few years have been working to identify and isolate the natural control agents that might be used in lieu of toxic chemical pesticides harmful to farmers and consumers alike.

Nigeria's International Institute for Tropical Agriculture has identified a certain species of wasp as a biological control for mealy bugs, one of cassava's main pests (*Special Report 114*, p. 17). Researchers at the Biological Control Program in Cotonou, Benin, have joined in making major advancements in destroying the mealy bug and the cassava green mite. Both parasites were accidentally introduced to Africa in the early 1970s and spread rapidly. The mealy bug has been brought under control over much of the continent and crop losses have been reduced drastically (Horst, 1990, p. 33).

From the late 1980s, biotechnologists have been working on maize and cassava equipped with the crops' built-in insecticides. They have already developed a streak-resistant corn. On the other hand, several species of chrysanthemum serve as essential components in the manufacture of pyrethrin, a widely-used modern insecticide (*World Development Forum*, 1989, p. 1). In Kenya, the Institute of Organic Farming has devised a cheap and effective means of ridding animals of ticks that consists of a tobacco solution mixed with soap which together forms a strong natural poison ("Ticks," 1991, p. 12). In Nigeria, derivatives of the common Neem tree have been found to repel 123 species of insects ("Neem," 1988, p. 14).

PROTECTING AFRICA'S GENETIC RESOURCES

Since the early colonial period, farmers have abandoned hardy indigenous food crops for the more easily cultivated and more profitable imported ones. Colonial administrators failed to appreciate the wide diversity of plants used for medicines, fuel, and shelter.

Today it is widely believed that Africa's dwindling tropical rainforests will provide the raw material for future crop species and the genes of plants that may be utilized in the twenty first century and beyond to develop medicines. Many African countries now recognize the need to build national warehouses for seeds and bits of DNA. A gene bank was established in 1987 in Nigeria by the International Institute of Tropical Agriculture in an effort to save West Africa's genetic resources of food crop species (Juma, 1989). A large volume of genetic material, including the wild ancestral species of modern sorghum, millet, pasture crops, and rice has also been collected in Kenya by the International Board for Plant Genetic Resources (Koffi, 1994, p. 32). At the international level, the Food and Agricultural Organization (FAO) has gathered information from African countries on breeds of seven animal species—the donkey, buffalo, cow, goat, horse, pig, and sheep—as part of a global data bank for animal genetic resources established to conserve endangered breeds (Food and Agricultural Organization, 1990, p. 1).

MEDICAL BREAKTHROUGHS

Wetlands in Africa have been grossly underutilized for food production largely because they have been the sources of such diseases as malaria, river blindness, and bilharzia. Consequently, wetlands have not been intensively farmed in tropical Africa but may soon be used when these key health hazards are removed and suitable technologies are put into place. Recent medical breakthroughs in our understanding of certain physically debilitating diseases may have a dramatic impact on public health and labor productivity.

Malaria is still considered Africa's number one killer, even with the growing rate of death from AIDS. According to the World Health Organization

(WHO) estimates, approximately 275 million people in sub-Saharan Africa carry the malaria parasite. Of the approximately 100 million new cases of malaria worldwide each year, about 90 percent are in Africa. But tentative test results of a new vaccine in the Gambia and Tanzania in 1993–1994 hold promise that malaria may at last be substantially reduced or even eradicated. Developed at the Medical Research Council in the Gambia, this chemically engineered vaccine increases the immune defenses, especially in children (French, 1994, p. A17). In Togo, biogeneticists are attempting to redesign the mosquito altogether. They have been working on new generations of malaria-resistant insects in order to block the spread of malaria, in other words to render the mosquito incapable of disease transmission by making it resistant to certain organisms (Harsch, 1992, p. 15).

Biotechnologists have also concocted a new drug against sleeping sickness, called Ornidyl. Sleeping sickness, a major health hazard that reduces worker productivity and severely impedes economic development, is caused by trypanosome parasites transmitted by the tse tse fly. Approximately 250,000 Africans are infected annually and more than 50 million are at risk. In cattle, the disease is called nagana and it can devastate entire herds in very short order. Much useful work is underway at the Uganda Trypanosomiasis Research Organization in Kampala and at the International Laboratory for Research on Animal Diseases in Nairobi, Kenya (Sjoerdsma & Schechter, 1985, p. 212).

Breakthroughs have also been achieved by Africans in the conquest of rinderpest. Not long ago, an Ethiopian microbiologist Tilahun Yilma came up with a genetically-engineered vaccine to eradicate rinderpest, a highly contagious viral disease that affects cattle. With a mortality rate of 95 percent, the virus is considered the biggest killer of cattle in Africa. Dr. Yilma has genetically-engineered a vaccine that is heat stable and easy to produce and administer. Farmers can administer the vaccine by making a simple scratch, and because the virus proliferates in skin cells they can grow their own supply by inoculating an animal and scraping the mark once a scab has developed. This procedure can yield 250,000 doses from a single animal. Rinderpest eradication promises to vastly increase beef and dairy production and to generate enough surplus for export to such livestock-deficit countries as Saudi Arabia. It would also enable the import of new breeds of high-output cattle from temperate countries (Fitzgerald, 1994, pp. 30–33).

Scientists also seem to be winning the battle against river blindness. This disease, caused by filarial worms that are spread by black fly bites, keeps people away from fertile lands around rivers and lakes. In 1988, a drug called Ivermectin was developed by Merck pharmaceutical company and enjoys a nearly 100 percent rate of effectiveness. It is anticipated that in time sleeping sickness will disappear entirely from the continent (Erinosho, 1990/1991, p. 21). In the fight against drug addiction, the rainforests of eastern Nigeria have offered a shrub which has been used to produce a drug called Ibogaine. Tests on Ibogaine suggest that it can reduce the incidence of relapse in a high proportion of drug abusers ("Ibogaine," 1990, p.17).

Finally, in a narrow forest strip along the northern coast of Madagascar grows the harmless-looking Rosy Periwinkle. From its delicate petals, researchers have synthesized a powerful anti-cancer drug that sends children suffering from certain forms of leukemia into long-term remission (Foster, 1990, pp. 35–36).

REGENERATIVE AGRICULTURE

Many constraints to intensive food-crop production in tropical Africa are related to soils. Fortunately we are seeing a proliferation of short and long-term research into soil and water resources and their management. Soil-management technologies suggest that higher food crop production can be achieved without degrading the soil and the environment.

Regenerative agriculture is a term used to describe efforts underway to restore productivity to lands that have been degraded through drought, deforestation, overutilization, or general abuse by human agency. Tropical Africa has perhaps the worst erosion problem on arable land in the world. New attention is being directed to bringing large tracts of semi-arid land into sustainable production. There is a growing appreciation that traditional agriculture is remarkably compatible with the ecological environment. In rural Kenya, where lands have been terraced and planted with trees, Africans have shown that land degradation can be arrested and even reversed (Makokha, 1987).

Africans are returning to pre-colonial traditions of land management practices that once worked well for their ancestors as they coped with long dry periods. One of the most revolutionary forms of soil-conserving agro-forestry for Africa is alley cropping, which involves growing food crops between hedgerows of leguminous trees that fix nitrogen in their roots. The technique permits continuous farming of the same plot while conserving soil and avoiding reliance on chemical fertilizers which most peasant farmers cannot afford (Ephson, 1990, p. 37).

The International Fund for Agricultural Development (IFAD), based in Rome, is following a new approach: looking for solutions in farmers' fields not simply in research labs (Rensberger, 1993, p. 10). Financially pressed donors and their recipients are searching for technologies that require little cost and minimal training.

In the West African sahel, an international consortium of Canadians, Americans and United Nations officials is joining Africans in reversing the southward movement of the Sahara by constructing windbreaks (or green barricades) of Australian pines. We now see the growth of a forest where a decade ago there was only sand. Other exotic tree varieties have been used in the sahelian region to stabilize sand dunes. The new technique called "deep-planting" utilizes the moisture that lies deep below the surface (Spoor, 1990, p. 20).

Great strides in water conservation have also been made in recent years by modifying traditional technologies. Farmers in Burkina Faso are using contour stone bunds—which are parallel lines of stones laid across the land's contour—to slow the flow of rainwater, allowing more to percolate into the ground. Ridge cropping, combined with wooden or stone cross ties in the furrows, has evolved over the last decade into an integral component of subsistence farming. Farmers have reported substantial increases in maize output under this tied-ridge system (Ouattara, 1990, p. 49). On the southern slopes of Mount Kilimanjaro, the Chagga farmers have returned to an impressive system of water management using tiny channels to irrigate their terraced hillsides. In doing so, they have been able to expand their cash earnings from irrigated vegetable gardening (Grove, 1993, p. 435).

Excessively irrigated farmland can become extremely salty. Sudan's Ministry of Agriculture has developed a new way to fight the problem of growing salinity in irrigated areas using a substance known as biological fertilizer, which is a mix of certain kinds of animal wastes and plant residues that have the ability to absorb salt and remove it from the water (Nduru, 1990, p. 26).

In tropical Africa, progress has been made not only in conserving water but in its delivery. Using sunlight to generate electricity is a dream of the pollution conscious. Now one form of solar power—the photovoltaic cell—is blossoming into a substantial tool for agriculture. Photovoltaic cells are valuable sources of electricity in areas too remote to be connected to a power grid. Photovoltaic technology could transform agriculture in Africa by providing power for simple pumping installations. Africans and the British have worked together to devise a mobile solar power and water pump/purification unit. In West Africa, the units are able to pump and purify up to 2,000 gallons of water daily (Dine, 1993, p. 40). Contaminated water has been a persistent public health problem in much of Africa and contributes to a wide range of digestive maladies.

SAVING THE OLD FORESTS BY CREATING NEW ONES

Deforestation has become a serious long-term threat to Africa due to the need of families to secure cheap fuel for cooking. Africans rely heavily on woodfuels for most of their energy requirements, despite the accelerating loss of forestlands. The development of more efficient, energy-saving, and inexpensive biomass cookstoves has alleviated local pressure on wood resources. Fuelwood accounts for about 80 percent of the energy needs of sub-Saharan Africa. Improved biomass stoves are a stepping stone between traditional stoves and modern fuels. Successful stove programs have been underway in Somalia, Kenya, and Rwanda. These innovative stoves can free firewood-collecting hands to work on crops in the fields. The fuel-efficient stoves are also realizing great savings in wood consumption. A new Somali ceramic stove consumes half the amount of wood of the conventional varieties because it is small, simple, and chimneyless. Only 5 percent of the heat energy of open fires is transferred to cooking pots. The biomass stove is a bridge technology between open cooking fires and kerosene and propane gas appliances, most of which remain beyond the reach of rural populations (Barnes, Openshaw, van der Plas, & Smith, 1994).

Wood stoves nevertheless need wood and efforts have been underway to protect traditional tall forests from destruction. There have been massive tree-planting programs and new fast-growing tree varieties. The Ghanaian government has encouraged rural communities to establish village wood lots where trees are grown exclusively for fuel and building materials. In so doing, valuable old forests can be preserved or at least the rate of cutting can be reduced (Ephson, 1990, p. 37). In Kenya, a group of local biologists formed a grassroots greenbelt movement and have planted over ten million trees since 1977. They have established more than 1,500 tree nurseries and have introduced tree-planting programs into more than 3,000 schools (Topouzis, 1990, p. 44). Lesotho, a country deeply eroded and denuded, has also been successful with a national tree-planting program. Forests are returning to a land bereft of trees for over a century and a half. In Uganda, where indigenous hardwoods take more than two centuries to mature, the newly-introduced Markhamia lotea trees, planted by the World Wildlife Fund (WWF) using the skills of village volunteers, can be harvested within a half decade. The alternative trees should make people less inclined to fell the hardwoods, which continue to provide a vital canopy for the fragile soils beneath them.

COMING TO TERMS WITH DROUGHT

Scientists have found a very close correlation between temperature cycles in the eastern equatorial Pacific ocean and rainfall and grain production in southern Africa. It was recently discovered that sea surface temperatures trigger long-distance atmospheric chain reactions that can influence weather conditions in other regions of the world. An understanding of this complicated process now enables planners to better anticipate and cope with droughts.

Working together, Zimbabwean and U.S. scientists have established a statistical correlation between the temperature cycle of the Pacific and rainfall patterns in Zimbabwe. As a result, they can better forecast the yields of maize, a food crop critical to the Zimbabwean economy. Maize needs moisture during inflorescence for the pollen to stick to the tassels. Working in synergy, an agricultural economist, an oceanographer, and a climatologist enabled African governments to develop effective early warning systems and thus anticipate and minimize the damage of drought (Cane, Eshel & Buckland, 1994, p. 204).

This enhanced ability in weather forecasting has already become a powerful new tool for agricultural planners as well as for farmers and relief agencies. Farmers can henceforth be warned in advance and prepare for impending dry weather by planting more drought resistant strains of crops, storing food, or conserving water for irrigation. In the severe drought of 1991–1992, the Southern Africa Development Community (SADC), a regional grouping of nations, established an extremely effective early warning system that coordinated the battle against an impending dry spell. Regional transport and communications were coordinated along with the international community. The strategy worked very well during southern Africa's worst drought of the century. The affected countries had time to work with international donor and relief agencies to deal with the impending problem and a catastrophe was averted (Collins, 1993).

CONCLUSION

It would be naive to conclude from this positive assessment that biotechnology is Africa's golden passport to future food security and economic advancement. Indeed, biotechnology is a double-edged sword, full of potential and full of danger. Biotechnology could reorder the world economy with Africa potentially the biggest loser. Left entirely in the hands of the industrialized West, it could undermine global markets for African commodities through the injection of less costly synthetic substitutes. Genetic engineering could also go awry and produce deadly new pathogens that could be used in biological warfare. Biotechnology could also spark a new kind of imperialism, where nations rob Africa of its genetic resources for their own gain.

These threats are very real but they must be weighed against the tremendous hope biotechnology offers Africa. Biotechnology provides substantial opportunities for continued improvement in the continent's food and health needs. It offers the prospect of higher yielding, longer lasting, and more drought resistant food and cash crops. Biotechnology may also be a means of developing essential new medicines to rid Africa of its debilitating scourges and to provide alternatives to the miracle drugs of our time that are fast losing their potencies. Africa is the custodian of much of the world's surviving genetic diversity, without which biotechnology could not go forward.

Africans are in the process of rediscovering their agricultural potential. They are re-examining what worked best in the pre-colonial past and have begun to harness the practices of earlier generations to the products of modern biotechnology.

In the final analysis, the full potential of biotechnology for African advancement will not be realized unless Africans direct more attention to the development of science in their own educational institutions. A higher level of scientific and technological capability is required in order to ensure that the beneficial results of biotechnology will accrue to Africa as well as to the industrialized West.

REFERENCES

Barnes, D.F., Openshaw, K., Plas, R. V. and Smith, K. (1994). What makes people cook with improved biomass stoves? *Technical Paper 242*. Washington, DC: World Bank.

Bennetzen, J. (1993). Grasses as a single genetic system. *Trends in Genetics, 9* (8), 259–60.

Brown, M. B. and Tiffen, P. (1992). *Short changed: Africa and world trade*. London: Pluto.

Bryant, E. (1993, October). Super-cassava. *African Farmer,* pp. 49–50.

Cane, M. A., Eshel, G. and Buckland, R. W. (1994, July 21). Forecasting Zimbabwean maize yield. *Nature 370*, p. 204.

Collins, C. (1993, August 3). Famine defeated. *Africa Recovery 9*, pp. 2–4.

Dine, R. (1993, October). Mobile solar unit is made for Africa. *African Business*, p. 40.

Ephson, B. (1990, November). Regenerative agriculture benefits Ghanaian farmers. *African Farmer 5*, pp. 37.

Erinosho, O. (1990/91). Health care and health care services in Nigeria. *Advance: Journal of the African Development Foundation, 3*, 19–23.

Fitzgerald, M. A. (1994). Man with a mission. *New Scientist, 142*, 30–33.

Food and Agricultural Organization. (1990, January/February). *FAO at Work.* Washington, DC: United Nations.

Foster, A. (1990, April). Farmers play vital role in development for nature swap. *African Farmer 3*, pp. 35–36.

French, H. W. (1994, September 22). Scientists say tests show new vaccine may prevent malaria. *New York Times,* p. A17.

Grove, A. (1993). Water use by the Chagga on Kilimanjaro. *African Affairs, 92* (368), 431–449.

Harsch, E. (1992, December). New push against malaria in Africa. *Africa Recovery,* pp. 14–16.

Horst, S. (1990, November). Building safe defenses against destructive bugs. *African Farmer,* p. 33.

Ibogaine research (1990). *Occasional Paper*, Ibadan University Hospital. *72*(14), 17.

Juma, C. (1989). *The gene hunters: biotechnology and the scramble for seeds.* Princeton: Princeton University Press.

Kesseba, A. (1989, October). Scientists till the soil of new food frontier. *The World Paper*, p. 360.

Koffi, J. (1994, January). Kenya scores success in environmental recovery. *African Business*, p. 32.

Makokha, J. (1987, March). The district focus in Kenya: The need for participatory organisations of the rural Poor. Paper of Kenya Economics Association. Nairobi, Kenya.

Nduru, M. (1990, July). Halting the salt that kills crops. *African Farmer*, pp. 22–26.

Neem trees and their uses. (1986). *Occasional Paper no. 77,* Ibadan University Hospital, p. 14.

Ouattara, S. (1990, April). Burkina Faso: An oasis of hope in the sahel. *African Farmer*, pp.49–50.

Pfizer (1993). *Annual Report.* New York: Pfizer, p. 24.

Rensberger, B. (1993, November 3). Ancient methods to save soil. *Washington Post*, p.10.

Schultes, R.E. (1981). *The winged bean: A high protein crop for the tropics,* 2nd ed. Washington, DC: National Academy.

Sjoerdsma, A. and Schechter, P. J. (1985, July). New drugs for sleeping sickness. *Journal of Clinical Pharmacology and Therapeutics*, pp. 211–24.

Special Report 114. (1992). Jos: International Institute of Tropical Agriculture, pp. 16–19.

Spoor, C. (1990, April). Finding buried water. *African Farmer*, p. 20.

Ticks and tobacco solutions. (1991, December). *African Farmer*, p. 12.

Topouzis, D. (1990, October/December). Kenya women fight deforestation. *Africa Recovery*, p. 44.

U.S. Department of State (1985, April). Africa: The potential for higher food production. *Special Report no. 125.* Washington, DC: U.S. Department of State.

Wainaina, S. (1991, December). Kenya sees bright future for the camel in Africa. *African Farmer*, p. 35.

Warner, P. J. (1992). Microbes in agriculture. *Science Progress, 76*, 94.

Washington Post (1994, July 27). Corn undergoing dramatic transformations, p. A3.

World Development Forum. (1989, February), p. 1.

Zymelman, M. (1990). Science, education and development in sub-Saharan Africa. *Technical Paper no. 124*. Washington, DC: World Bank.

Part VI

A VISION OF THE FUTURE

11

Afrenaissance: Struggles of Hope in Post-Colonial Africa

Ali A. Mazrui

Good news from Africa is often lost because reporters focus on the wrongs the governments perpetrate rather than the precise nature of the people's reactions and response. In September 1994 the world press rightly reported on the banning in Nigeria of such newspapers as *The Guardian* and the *Concord* family of publications. The international press also reported on the new decrees about detention without trial and the effort to put military decisions beyond judicial challenge and review. All this reporting of the bad news from Nigeria was quite justified. What was not reported was the continuing defiance by much of what remained of the Nigerian press. In spite of the threat of detention without trial, many reporters and feature writers continued to stick their necks out on important political issues, often critical of the government of General Sani Abacha.

In September 1994 I attended a conference of the youth of Nigeria titled "Nigeria in the 21st century." The conference was sponsored by the Yakubu Gowon Center (named after a former military head of state) and the conference was opened by another military head of state at the time, General Sani Abacha. Nevertheless, the conference included papers strongly critical of both the excesses of military regimes and the corruption and anti-democratic cynicism of many Nigerian civilian politicians. Clearly debate and independent expression were alive and well in Nigeria, in spite of draconian military decrees (Rotimi and Ihonvbere, 1994).

These kinds of contradictions in Nigeria signify the reality of *Afrenaissance.* We define *Afrenaissance* as living evidence of Africa's self-renewal, living testimony that the human spirit is reasserting itself among the African people. But against what odds?

BETWEEN ETHNICITY AND LABOR

In places such as Rwanda we have witnessed not just the decline of the nation-state but its complete collapse. We have also seen Western hegemony at work. In Rwanda the West had the power to decide in 1994 whether to save only Western nationals in the mayhem and let the Hutu and the Tutsi carry out their collective suicide pact. The West, in 1994, also had the power to get the UN Security Council to play the pendulum game—one day calling on the United Nations forces to withdraw from Rwanda and let the violence take its natural course, and the next week calling on the United Nations to increase its presence. There was also the power of Elysee Palace in France, which ranged from being the main supplier of Hutu military destructiveness to being the main protector of Tutsi on the run. The French presidential palace played goddess of death on one day, angel of life the next.

Why did not Africans intervene to end the genocide in Rwanda? They did. "Africa" took the form of the Rwanda Patriotic Front, which, in the face of immense ethnic provocation, managed to limit its goal to conquest rather than carnage, dreaming of restoring order rather than avenging terror. The Patriotic Front's discipline could not be expected to last unconditionally. If justice was not done to genocidal criminals, Tutsi self-restraint could have eroded. But in 1994 the Patriotic Front showed more self-restraint under provocation in several months of stress than any other band of soldiers anywhere else in the world in similar circumstances.

Western hegemony is not omnipotent. In 1961 John F. Kennedy launched an invasion of Cuba by Cuban exiles on the Bay of Pigs. The invasion, intended mainly to serve U.S. interests, was a complete fiasco. In 1994 President Yoweri Museveni of Uganda helped Rwandan exiles launch their own "Bay of Pigs." It was successful. For once an African initiative has been more triumphant than its Western-sponsored predecessor. This was another sign of Afrenaissance.

When Moshood Abiola won the Nigerian presidential election in June 1993, he was the first southerner (in Nigeria's political parlance) to be *elected* executive head of state in the country's post-colonial history. In previous elections Nigerians brought northern Muslims into power. In June 1993 they still brought back a Muslim victor—but this time he was from the South. This was a step forward in Nigeria politics. Did the military government of Ibrahim Babangida nullify the elections for those ethnic and regional reasons? Was Abiola—who died mysteriously in 1998 on the eve of his probable release from prison—denied the presidency because he was Yoruba rather than Hausa/Fulani?

On being denied the presidency, Abiola's biggest mistake of 1993 was in exaggerating the scale of Western hegemony in the world. Instead of staying at home to fight it out, he traveled to the capitals of the Western world to seek international help in the fight against the military government in Nigeria. Washington and London made small gestures. Abiola damaged himself seriously at home.

In 1994, a few weeks before he resumed his struggle for the presidency, Abiola was in Washington again, but this time he knew that the main battlefield had to be Nigeria. He telephoned my home and missed me. With the help of my wife, he tracked me down at a hotel at Lincoln University in Pennsylvania and told me: "I am going to Nigeria to become president. I will see you at the inauguration!" I was startled. In a sense I am still startled, but that phone call was also a lesson to me in political courage. A few weeks later Abiola declared himself president of Nigeria at a rally in Lagos attended by thousands of people. He should have done so a year earlier. He had gambled in 1993 on Western hegemony and lost. But on matters of principle, "better late than never"!

Abiola's electoral support had been multi-ethnic. That also was a step forward in Nigeria, but his support in the fight against the military regime became heavily Yoruba and in the Western region. This was a relapse. However, the fight produced one remarkable phenomenon—the strike of the oil workers and their supporters in defense of Abiola. The strike became the most impressive utilization of labor power for democratic ends in the history of post-colonial Africa—regardless of whether or not the strike ultimately succeeded. Its capacity to sustain itself for many weeks and hold the nation's economy to ransom on an issue of national democratic principle had already earned the strike a place in post-colonial history. It was a sign of Afrenaissance. The strike was finally broken after its eighth week.

In Poland under communism in the 1980s the defiance of Solidarity as a trade union movement had been widely acclaimed in the Western world. The petro-labor strike in Nigeria in the 1990s in defense of democracy fired few imaginations in the Western world, in spite of the fact that Nigeria was several times the size of Poland in population, and was Africa's most populous country. Lech Walesa became a household name in the West, and he won the Nobel prize for Peace in 1983. Who knew the names of Nigeria's trade union leaders outside Nigeria? Was any of them likely to win the Nobel Prize for Peace?

The double standards of the World had persisted, distorted by Western hegemony. What passed for heroism in Europe (i.e., Solidarity in Poland) passed for instability in Africa (i.e., the strike of Nigeria's oil workers). What passed for martyrdom in Eastern Europe (harassment of Lech Walesa) became another example of African tyranny when African democrats were victimized. The Western focus was disproportionately on who was killing democracy in Africa, and inadequately focused on who was fighting for democracy on the continent. African villains got more coverage than African heroes.

PARADOXES OF AFRICA'S DEMOCRATIZATION

In countries such as Kenya, United States foreign policy has in the past been oriented toward the ambition of making democracy happen. What is perhaps more compelling is the need to make democracy endure. Americans have often been ready to spend a lot of money to undermine dictatorships in countries as far-flung as Ethiopia, Afghanistan, Angola, Cuba, and Haiti. But Americans have been less enthusiastic about spending money to stabilize fragile democracies in places such as Zambia, Malawi, or even Eastern Europe. A shift in emphasis may be needed in American foreign policy from undermining dictatorships to stabilizing democracies. Whole new strategies may need to be worked out to help democracies endure.

Economically this may mean helping new democracies be economically successful—and making a democracy as a system deliver some of the basic needs of the people. Places such as Malawi, Zambia, and South Africa may also need considerable help in training human power for administrative, parliamentary, party, executive, judicial, and related roles.

In Africa there is clear evidence that the people are tired of dictators. In one African country after another dictators have been removed. There is a genuine desire for democracy, but the capacity to democratize needs to be strengthened; democratic Afrenaissance needs to be consolidated. The democratic spirit may be willing, but the political flesh is weak. The United States can help democracy endure in Africa only in certain ways, but both Africa and the United States must first recognize this paradox of a strong African desire for democracy combined with the fragile African capability. We need to understand the malaise before we can help treat it.

In the course of the 1990s much of Africa continued to demonstrate this solid desire for democracy. What has yet to be tested fully is Africa's capacity to democratize. The desire for democracy has manifested itself in a variety of ways—from pro-democracy street demonstrations in Togo to the Muslim riots at the Kenyan coast, from the political tug-of-war between soldiers and civilians in Nigeria to the electoral fall of Hastings Banda's tyranny in Malawi. But are Africa's will and capacity to democratize strong enough? This capacity requires not only the urge to throw out one particular unpopular regime, but the determination to institutionalize legitimate political succession indefinitely. The will to democratize also requires readiness to permit an unpopular but legitimate government to govern without hindrance and complete its electoral term of office without being overthrown.

Ghana elected Kofi Busia in 1969, but the Ghanaian army overthrew him in 1972 before he completed his first electoral term. In 1979 Ghana elected Hilla Limann, and again the Ghanaian army overthrew him in 1981 before he completed his first term. What was even more astonishing, popular opinion in Ghana cheered both military coups. The Ghanaian will to democratize was not yet strong.

In 1992 a new situation arose. Flight Lieutenant Jerry Rawlings, formerly a dictator, resigned his military commission and stood for election as a civilian. He won, but his opponents disputed the results. Did he win because the register of voters was unfairly incomplete, because of actual vote-rigging during the election, or were the opposition parties simply bad losers? Whatever was the truth, it constituted evidence that Ghana's democracy was not fully consolidated.

The evidence that Jonas Savimbi and UNITA were bad losers in Angola is probably clearer. Former United States' protégé Jonas Savimbi has repeatedly plunged the country into renewed civil war rather than accept the results of presidential elections, which had gone against him, or abide by peace agreements he subsequently signed. Angola's will to democratize was not surprisingly weak. The country had been ruled by Portugal, one of the least democratic of all European colonial powers in Africa. In the last decade of colonial rule, Angola was more accustomed to multi-army politics than to multiparty politics. The democratic tradition in Portuguese-speaking Africa had either been destroyed by the whole Portuguese domineering style or never arrived because Portugal ceased to be part of the European mainstream, and its colonies were ignored or brutalized as a result. Mozambique suffered a comparable fate—more a multi-army experience than a multiparty tradition. The will to democratize is particularly weak in lusophone Africa.

When I was at the summit meeting of the Organization of African Unity in Dakar, Senegal, in June 1992, I was privileged to be introduced to Zambia's then new head of state, President Frederick Chiluba. He had learned a lot in a year. An important lesson for democracy is to not only be a good loser, but also to be a gracious winner. At first Chiluba's party was unable to grasp that point. The party seemed determined to humiliate former President Kenneth Kaunda, which seemed to be the clearest way of sending a message to presidential incumbents throughout Africa not to make "the same mistake as that made by Kenneth Kaunda"—that is, give in to democratic pressures.

It would be a disaster to send a message to African presidential incumbents that humiliation awaits them if they cease to be heads of state. Chiluba had a special responsibility to Africa as a whole to avoid sending such a message, and thanks partly to Chiluba's own good sense and partly to the advice of General Olusegun Obasanjo and other African elder statesmen, moves were made toward reconciliation between Chiluba and Kaunda. President Chiluba's highly publicized birthday greeting to former President Kaunda were particularly symbolic.

More needs to be done in Zambia and the rest of Africa to make former presidents feel honored and respected even when they are defeated. Nigeria is already on the way. General Yakubu Gowon is greatly respected, though there is considerable controversy with regard to former President Shehu Shagari (civilian) and former President Ibrahim Babangida (military). In the United States all former presidents are politely addressed as "Mr. President."

The will to democratize is sometimes compromised by an unwillingness to accept the triumphant ideology in a contest. This was most dramatically illustrated in Algeria. Despite the pluralistic experiment, the powers in Algeria were not ready to accept the triumph of the ideology of the Islamic Salvation Front. Elections were held and then aborted when the winner was not acceptable to those who were able to command military support to frustrate the democratic process. Algeria was plunged in a "no-win" situation as the democratic secularists sought to neutralize the majoritarian Islamists.

One of the major dilemmas confronting the new African democracies is whether economic liberalization should have preceded political liberalization. Should Africa have reduced the role of the state in the economy before Africa increased the role of the people in the political process? Would that have strengthened Africa's will to democratize (Apter & Rosberg, 1994)?

Experience in Asia seems to lend support to that strategy. South Korea embarked on vigorous capitalism before attempting liberal democratization. On the whole this approach has also been true of Taiwan, and of most members of the Association of South East Asian Nations (ASEAN). Successful Asian capitalist countries have pursued the legacy of Adam Smith before they paid attention to the legacy of Thomas Jefferson or the Jacobins.

An even more dramatic case is that of the People's Republic of China since Mao Tse Tung's death. The giant of Asia has been moving toward greater economic liberalization, while remaining impeccably opposed to political liberalization. The Tiananmen Square suppression of 1989 was a measure of opposition to political liberalization, but China's commitment to economic liberalization has remained unabated. The Chinese people have a desire for democracy, but not yet the will to force the aging elite to provide it.

On the other hand, Mikhail Gorbachev attempted political liberalization (*glasnost*) before any systematic economic liberalization (*perestroika*). The result in the former Soviet Union was rapid political disintegration without meaningful economic transformation. It was not only the Soviet empire of Eastern Europe that disintegrated in the wake of *glasnost*, it was also the bicontinental body politic of the Soviet Union itself. The Chinese are justified in seeing Gorbachev as the man who destroyed the USSR, the country which made him leader.

In Africa, Ghana has from time to time explicitly faced the dilemma as to which should have priority first—political recovery or economic recovery. Ghana's first post-independence president, Kwame Nkrumah, opted for the primacy of politics. He declared in the 1950s: Seek ye first the political kingdom and all else will be added unto it.

Did Nkrumah mean that political independence for an African country would bring "all else added unto it"? Or did he mean the liberation of the whole of Africa (including the Republic of South Africa) was a precondition for Africa's "political kingdom"? Or was his concept of "political kingdom" a combination of both Africa's liberation and Africa's unification into a united country on a continental scale?

Clearly political independence for individual African countries has never "added all else unto it." For individual African nations Nkrumah's emphasis on the state as the primary actor in Africa's development has been invalidated (Mazrui, 1994a). More than 30 years after Nkrumah's dictum of seek ye first the political kingdom, another Ghanaian leader drastically revised the dictum. Jerry Rawlings' politics from the mid-1980s seemed to be based on the dictum: Seek ye first the economic kingdom and all else will be added unto it. Rawlings—in spite of left wing rhetoric—submitted his Ghanaian government to the rigors of structural adjustment and related discipline of the World Bank and the International Monetary Fund (IMF).

It was not until 1992 that Jerry Rawlings' government, under both domestic and international pressures, at last conceded political democratization and implemented multiparty elections. Political liberalization eventually caught up with economic liberalization in the new Ghana. Was "the economic kingdom" about to demonstrate that it would have "all else added unto it"? The question is still wide open as to whether that approach in Ghana will be as vindicated as it has been in South Korea, Taiwan, South-East Asia, and possibly the People's Republic of China. Might Rawlings have negated Nkrumah's premise of the primacy of African politics? Or is there a third dictum which demands to be explored: Seek ye first the cultural kingdom and all else will be added unto it? (Mazrui, 1994a; World Bank, 1994).

The fate of both political and economic liberalization hinges on cultural variables which have too often been underestimated. We may need to grasp the cultural dimension before we can fully gauge the scale and durability of social change in Africa. Every constitution needs to be culturally viable. Every development project needs a cultural feasibility study.

However, history does not wait for historians; it has its own momentum. In many African countries political liberalization has been taking place without either a cultural feasibility study or an economic stock taking. Liberal democratic activism has been underway from Madagascar to Mali, from Marrakesh to Maputo, from Dar-es-Salaam to Dakar. The scale of activism has varied from country to country, but a liberal, pluralistic contagion has been spreading across the continent. Yet the question persists: Is it only a desire for democracy or is there a real will and capacity to democratize?

We have posed a wider scenario: "Seek ye first the cultural kingdom and all else will be added unto it." This third scenario requires understanding the cultural preconditions of both political will and economic competence. Africa needs planned democracy—preceded by a feasibility study to ascertain the cultural viability of any proposed economic or political system. Slogans such as "privatization" and "democratization" need to be accompanied by a cultural manual specific to the society or to the region. Africa's desire for democracy has reached significant proportions. Devising a culturally inspirational agenda for the future may transform that democratic desire into a powerful democratic will. Democratic Afrenaissance may at last find a firm foundation.

Most of the democratizing work has to be done by Africans themselves. The United States can try to save new democracies from becoming economic failures. The United States can also help transmit technical skills of election monitoring, party compromises, parliamentary effectiveness, administrative capacity building, and the training of journalists and reporters for a free press. The United States may also need to be more sensitive to the speed and sequence of economic liberalization in relation to political liberalization. The African democratic spirit is still willing even if the political flesh is still weak. In time a strong desire for democracy may indeed crystallize into a strong will and a capacity to democratize.

Africa needs to find a middle ground between democracy as a cause of conflict and democracy as a cure. Africa has to tilt the balance away from democracy as part of the problem (multiparty tribalism) to democracy as part of the solution. But how? There are general principles and values concerning conflict management in Africa that need to be addressed. There are also levels of conflict which need to be understood and contained or resolved; and third, there are methods and mechanisms for the prevention, containment, or resolution of conflict which may need to be created.

Both political pluralism in the sense of multiparty democracy and economic pluralism in the sense of private enterprise and a market economy involve competition. One persistent problem in African multiparty politics especially has been how to prevent political competition from disintegrating into political conflict. Africa has already sensed the need for independent monitors (usually foreign) at election time. The monitors have not always ensured absolutely free and fair elections, but it is likely that their presence and role have sometimes averted the disintegration of electoral competition into electoral conflict.

TOWARD MOTIVATING PEACEFUL SUCCESSION

The question is beginning to arise as to whether African democracies need to be monitored continuously rather than merely at election time, and whether the African Union (AU) should include mechanisms for monitoring at least performance in human rights, if not the democratic process as a whole.

In addition to preventing the escalation of competition to conflict, there are occasions when we need to prevent the escalation of conflict into large-scale civil violence. The causes may have something to do with the quality of governance, and not necessarily with the party system per se. Is Africa ready for a pan-African judicial system with an African court of justice? The climate of democratization in individual countries needs to be strengthened by pan-African mechanisms of accountability—from an AU ombudsman to methods of honoring heads of state who bow out of power gracefully and permit peaceful political succession.

As noted, in many African countries presidential power is a zero-sum game—either one is head of state or a figure of fun and derision, if not at risk for imprisonment or execution. This state of affairs encourages some heads of state to cling to power indefinitely because they have everything to lose by giving up the state house.

As we have argued earlier, it is imperative that conditions be created to honor African presidents who bow out of power. Some progress has been made in countries such as Nigeria and Tanzania, where there are examples of former heads of state who still enjoy great public respect. Zambia could have tried harder to honor former President Kenneth Kaunda instead of harassing him from time to time. Senegal's relationship with its former head of state, Lèopold Sedar Sènghor, is more subtle and more complex. The political succession was peaceful, but was the departing president adequately rewarded?

Former heads of state who step down from power voluntarily should, first, receive a handsome pension; second, retain the honorific title of "president" (the way the Americans allow their former presidents to retain the title); third, receive an allowance towards an effective bodyguard; and fourth, be a member of a new AU institution—a biannual meeting for former heads of state who were replaced from power democratically (a summit of elder statesmen). This will be a kind of pan-African house of lords or pan-African senate, which can make recommendations either to individual African governments, or to the summit of incumbent heads of state of the AU and, on some issues, make recommendations even to the United Nations.

Two of the main purposes of this pan-African senate of elder statesmen and stateswomen would be, first, to tap the experience of former heads of state at a time when they are no longer constrained by the exercise of power; and, second, to provide an additional pan-African acknowledgment and salute to those African heads of state who not only served their countries but also allowed themselves to be replaced from power democratically.

PEACE KEEPING UNDER PAX AFRICANA

There are occasions when competition cannot be stopped from degenerating into conflict, or conflict into civil violence. Indeed, civil violence may sometimes escalate to full-scale civil war. Is Africa ready to deal with civil wars within its own borders?

The end of the Cold War eliminated one of the factors that prolonged many African civil wars. Superpower rivalry used to encourage internecine African brutalities instead of sponsoring African reconciliation, but now the big powers are trying to pressure Angola's MPLA government and the UNITA rebel movement, for example, to reach genuine peace accords and abide by them. A few years ago the same external powers were arming UNITA and the MPLA to the teeth to continue their fratricide against each other. On the other hand, a more general collapse of law and order in some African countries (such as Somalia and Sierra Leone) is raising the specter of recolonization by outsiders and a new foreign trusteeship system. The clouds of a new (if more humane) imperialism are real.

Can Africa construct a peace-keeping force of its own? Can it set up a peace-making or even peace-enforcing machinery? There is indeed a need for a Pax Africana—an African peace established, enforced, and consolidated by Africans themselves. Military and diplomatic interventions by the Economic Community of West African States (ECOWAS) in dealing with the civil wars in Liberia and Sierra Leone were examples of what could be accomplished with Pax Africana.

However, it was during a stalemate in Liberia that anarchy descended on Somalia in the early 1990s. In a sense, Somalia suffered because of the stalemate in Liberia. While African states struggled to end the civil war in Liberia, what chance did they stand with the anarchy in Somalia? But as the United Nations has become over-stretched in other parts of the world, Africa may have to develop a new regional capability of its own in peace-keeping and peace-enforcement. The frustrating lessons of Liberia and Sierra Leone need to be studied more carefully in order to avoid repeating some of the errors of the past.

One question to be resolved is whether Africa should decentralize its peace-keeping machinery, sub-region by sub-region (East, West, North, South, and Central Africa). Under such a design, it was correct that the experiments in Liberia and Sierra Leone were undertaken by West African states under the aegis of ECOWAS. If there had been an African initiative for Somalia, should East African states have taken the lead?

The sub-regionalization of Africa's future peace-keeping machinery is certainly one strategy worth considering. The argument against it is that while neighboring countries understand each other's problems better, they sometimes develop sub-regional tensions. Using Ethiopian troops for peace-keeping in Somalia may be more risky than using Ghanaian troops. Ethiopian troops in Somalia may not always be neutral. On the other hand, it may be better to use Ethiopian troops for peace-keeping in Togo than Ghanaian troops. Sub-regionalization of peace-keeping must therefore be handled with care and critical sensitivity.

POSITIVE VALUE: SHORT MEMORY OF HATE

What cultural resources or traditional mores does Africa have that may be tapped to help contain or resolve conflict? Perhaps the most important is Africa's short memory of hate. Time and time again Africans have displayed a remarkable capacity to let bygones be bygones.

Jomo Kenyatta was imprisoned for years by the British and denounced by a British governor as a leader unto darkness and death. When he finally came out of detention to lead a newly emerging independent Kenya, Kenyatta became one of the country's leading anglophiles. He turned the foreign policy of independent

Kenya toward Britain and the West. His book *Suffering Without Bitterness* is testimony to the African capacity to forgive.

Nelson Mandela lost nearly three decades of his life, imprisoned under white South Africans' laws. He came out of prison in 1990 ready to forgive and forget, as long as apartheid ended. Mandela was even capable of feeling compassion for three white terrorists ostensibly fasting unto death in protest against their own white government. Mandela visited the white terrorists personally to beg them to eat. A short memory of hate indeed.

Ian Smith in the old Rhodesia unleashed the horrors of a Unilateral Declaration of Independence (UDI) on his people, and was therefore instrumental in initiating a war in which thousands upon thousands of Africans were killed. When the war ended, and Zimbabwe became independent in 1980, this architect of the blood-letting UDI was not put before the equivalent of Nuremberg trial, but became a member of parliament of the new Zimbabwe and proceeded to abuse the new black government of the day. In some parts of the world Ian Smith would have been assassinated or executed as a war criminal. In Africa he was a beneficiary of black Africa's short memory of hate. (This inclination to forgive has been overshadowed by the recent violence of former nationalist fighters who are taking over white Africans' farms by force in their quest for ownership rights.)

As the Nigerian civil war was coming to an end, many people expected rivers of blood in the former Biafra—vengeance perpetrated by the victorious federal side. The Pope in the Vatican was among those who were deeply concerned about the likely fate of the Catholics in Iboland. Will the euphoria of military victory result in an orgy of blood-letting? None of these fears were borne out by what happened. Unruly as Nigerians sometimes are, they found self-restraint at the supreme moment of the conclusion of a civil war. There was no free for all orgy of revenge and no Nuremberg trials. On that occasion, Nigerians were tested for moral discipline, and they were not found wanting. They were the carriers of Africa's short memory of hate.

While African civil wars and fratricides get considerable global attention from the international media, Africa's rapid moments of forgiveness are not a matter of headlines. International news coverage suggests that to kill is news, to forgive is not. A shot is a story, an embrace seldom is. Africa's short memory of hate is not, therefore, celebrated internationally. Nevertheless, that limited memory of animosity is an important cultural resource, a valuable traditional more, for future conflict-resolution. Africa needs to find ways of tapping it to enhance the consolidation of Pax Africana.

POSITIVE VALUES: ECUMENICALISM AND THE ELDER TRADITION

Another cultural resource and value that needs to be tapped is the elder tradition in Africa. We have already proposed the establishment of a pan-African senate of elders consisting of former presidents who stepped down voluntarily to give way to democratically selected successors (e.g., Kenneth Kaunda, Lèopold Sènghor, Julius K. Nyerere). These same elders could also be requested from time to time to serve as mediators or conciliators in certain situations. General Olusegun Obasanjo, Nigeria's military head of state from 1976 to 1979, performed mediating tasks outside the purview of the Organization of African Unity before his imprisonment by General Abacha's regime in 1995 and his later election as Nigeria's civilian president in 1999. Former U.S. President Jimmy Carter has also engaged in comparable mediating roles from time to time. If a pan-African senate of elders is institutionalized, it could provide a pool of roving conciliators and mediators. Africa's respect for elders could be tapped to help the success of such roles.

The third cultural resource that may be tapped for conflict prevention, containment, or resolution is Africa's ecumenical spirit. On the whole, religion in Africa is divisive only when it reinforces other divisions. In Nigeria almost all Hausa are Muslim, almost all Igbo are Christian, and the Yoruba are split between the two religions. Thus, Christianity tends to reinforce Igbo identity and Islam tends to reinforce Hausa identity. The imported religions help to sharpen preexisting differences. In Sudan, Islam and Christianity also reinforce other immense ethnocultural differences between the North and the South of the country. The two Semitic or Abrahamic religions (Christianity and Islam) are additional catalysts of regional discord, rather than the main causes of differentiation.

On the other hand, Senegal is over 80 percent Muslim, and yet for 20 years the country had a Roman Catholic president (Lèopold Sedar Sènghor) without Islamic fundamentalist riots in the streets of Dakar protesting "Christian minority domination." Senegal was at once one of the most open African countries and one of the most ecumenical. When Lèopold Sènghor voluntarily retired from power, Senegal finally had a Muslim president, though even then the first lady of Senegal was a Roman Catholic. It is a record of presidential ecumenicalism which has no equivalent in the Western world or anywhere else outside Africa. The United States has yet to experiment with a Jewish president, let alone a Muslim one.

In Tanzania the Muslim half of the population had a better record of accepting the leadership of Julius K. Nyerere (a Roman Catholic) than the Christian population has had accepting the leadership of Ali Hassan Mwinyi (a Muslim). But, on balance, Tanzanians are still a good illustration of the ecumenical spirit at work in Africa. Indeed, many African families in many parts of the continent are themselves multireligious without breaking up—a Sunni father, a Unitarian mother, a Catholic uncle, an Ahmadiyya son, an Anglican daughter, perhaps all of them with a partial allegiance to African traditional religion.

If the ecumenical spirit is indeed one of Africa's cultural resources, how can it be tapped in the struggle for containment or conflict resolution in the continent? There has been an under utilization of religious leaders in Africa even in those conflicts perceived as being, in part, sectarian. Until the visit of Pope Paul to the Sudan in 1993, there had in fact been limited participation by major religious leaders in finding a solution to the Sudanese problem. The tendency was to look almost exclusively toward such political mediators as President Ibrahim Babangida of Nigeria, President Yoweri Museveni of Uganda, and President Jimmy Carter of the United States.

Should there be more systematic use of Africa's religious wisdom in the quest for Africa's sectarian problems? Archbishop Desmond Tutu has been used to help solve political problems as far away as Thailand and Burma, but has his wisdom been used adequately to help solve problems in the Sudan and Nigeria? Is Desmond Tutu a prophet in Asia, but not in his own Africa?

The proposed pan-African senate of elders should include a few select religious figures from Africa—figures such as Bishop Tutu, or the Sheikh of Al-Azhar University (the most senior representative of the Islamic wing of this ancient university), the most senior Roman Catholic Cardinal in Africa, the head of the Orthodox Church of Ethiopia, and a distinguished exponent of the indigenous religions of Africa. Just as the House of Lords in Britain includes bishops as members, the pan-African senate of elders should include a selection of Africa's religious leaders.

POSITIVE VALUE: WOMEN BETWEEN WAR AND PEACE

The fourth cultural resource or traditional more from Africa that may be relevant for conflict resolution is the traditional involvement of women in consultations about war and peace. The liberation wars in Southern Africa in the second half of the twentieth century produced many strong women, both among fighters in the bush and political activists. In spite of her subsequent legal and political setbacks, Nomzamo Winnifred Mandela of South Africa was often a beacon of hope and struggle, especially in the 1980s. A less controversial but equally militant activist was Nohtsikelelo Albertina Sisulu, who was also subjected by the apartheid regime to recurrent harassment, banning, and detention. (For some of the wider regional issues concerning women in Southern Africa see, for example, Armstrong, 1992).

In recent cultural history, indigenous women have been known to become soldier-priestesses, leading men into battle. In the 1980s, the government of Yoweri Museveni in Uganda had to contend with the rebellion led by Alice Lakwena, who had convinced tough Acholi warriors that she had religious powers which would protect them from bullets. Alice Lakwena was briefly the Joan of Arc of the Acholi—and, like Joan of Arc, Alice was defeated. But sociologically it was significant that a woman could convince members of one of the most martial "tribes" in Uganda to follow her into battle. In the 1960s it was Alice Lenshina in Zambia who was leader-priestess in defiance of the government of President Kenneth Kaunda. She and her Lumpa Church were militarily defeated. But at least as significant was the sociology that made female militant leadership of men possible.

Are there ways of tapping African women's wisdom toward prevention, containment, or resolution of conflict? At this stage we can only make a start. So far we have mentioned two categories of members of the proposed pan-African senate of elders—former presidents (democratically replaced) and major religious leaders. The third category of members should be outstanding African women, such as Angie Brooks of Liberia, who in 1969 served as one of the first women presidents of the United Nations' General Assembly. How such women will be chosen to serve on the pan-African senate of elders is something to be thought through. One possibility would be for the summit of the African Union heads of state to create an electoral college consisting exclusively of independent women, who would choose which women would serve on the senate of elders. Members of the electoral college would themselves be disqualified from being elected to the senate.

CONCLUSION

If prevention is better than cure, it follows that reducing the risk of conflict is better than strengthening methods of resolving conflict. Africa needs to find its way toward (1) conflict prevention, (2) conflict containment, and (3) conflict resolution.

Understanding some of the causes of conflict in Africa becomes an imperative. Those causes do include ethnic rivalry, uncertain processes of political succession, the uncertain equilibrium in civil-military relations, political repression, and economic deprivation. Many of those causes are beyond the capacities of any regional organization such as the AU, and some of the problems may take decades to resolve. In this chapter we have limited ourselves to adjustments which may be feasible even at the level of pan-African initiative or through the actions of the AU.

In spite of some sad lessons of the role of the Economic Community of West African States (ECOWAS) in Liberia and Sierra Leone, Africa will need a peace-keeping and peace-enforcing force of its own. The threat of re-colonization by others may not be idle talk. If there are a few more political collapses like that of Somalia and Liberia, a trusteeship system of a kind may return to Africa. (For an ominous article on recolonization, see Johnson (1993, April 18). Disguised re-colonization is indeed possible unless Africa can establish its own machinery of Pax Africana, an African peace maintained and consolidated by Africans themselves. It can be self-pacification, or even self-colonization (Mazrui, 1994b). Can this best be done by sub-regional peace-keeping forces? These are details to be worked out.

Peace-keeping is not enough, however. Africa needs to democratize itself and stabilize its democracies. African mechanisms for monitoring each country's performance in human rights is an imperative.

Africa also needs to find ways of rewarding good rulers, not simply ways of harassing bad rulers. At the minimum, African presidents who hand over government to democratically elected governments should be handsomely pensioned and appropriately honored.

There are African values and mores which are relevant for conflict prevention, containment, or resolution. An effort should be made to identify those values and ascertain how to tap them for practical purposes. Among those creative values are Africa's short memory of hate, traditional allegiance to elders, leaning toward the ecumenical spirit, and traditional readiness to involve women in decisions about war and peace.

The struggle must continue to prevent peaceful political competition from becoming heated political conflict, and prevent the latter from escalating into large-scale violence. There have been too many times when war itself has cast its shadow on African societies. Africa will not find durable peace until democracy is part of the solution. For too long in post-colonial Africa, democracy has been part of the problem.

An African capacity for self-participation needs to be married to an African dedication to self-democratization. Against the background of such a marriage, African swords may in time be turned into ploughshares, African spears transformed into elongated spades. The continent's self-renewal has gotten underway. It will all add up to the miracle of Afrenaissance at long last.

REFERENCES

Apter, D. E. & Rosberg, C. G. (Eds.). (1994). *Political development and the new realism in sub-Saharan Africa.* Charlottesville: University of Virginia Press.

Armstrong, A. (1992). *Struggling over scarce resources: Women and maintenance in Southern Africa.* Harare: University of Zimbabwe Press.

Johnson, P. (1993, April 18). Colonialism's back—not a moment too soon. *New York Times Magazine*, pp. 22–23, 43–44.

Mazrui, A. A.(1994a). Development in a multicultural context: Trends and tensions. In I. Serageldin & J. Taboroff (Eds.), *Culture and development in Africa* (pp. 127–136). Washington, DC: World Bank.

— (1994b). The message of Rwanda: Recolonizing Africa? *New Perspectives Quarterly*, *2*(4), 18–20.

Rotimi, A. O. & Ihonvbere, J. O.(1994). Democratic impasse: Remilitarisation in Nigeria. *Third World Quarterly*, *15*(4), 669–690.

World Bank. (1994). *Adjustment in Africa: Reforms, results and the road ahead.* Washington, DC: World Bank.

Index

About the Contributors

NJOKU E. AWA is Associate Professor in the Department of Communications, Cornell University. Dr. Awa's areas of specialization include development communication, interpersonal/intercultural communication, and persuasion and attitude change. He is former director of graduate studies in communication at Cornell.

H. SYLVAIN BOKO is Assistant Professor in the Department of Economics at Wake Forest University. He is the author of several articles on economic policy-making and reform in Africa. His book *Decentralization and Reform in Africa* is forthcoming from Kluwer Academic Publishers.

PETER P. EKEH, a political sociologist, is Professor of African American Studies at the State University of New York at Buffalo. He is the author of *Social Exchange Theory: The Two Traditions* (1974), along with numerous journal articles and book chapters.

RICHARD W. HULL is Professor of African History and Civilization at New York University and president of the New York African Studies Association. Dr. Hull's publications include *American Enterprise in South Africa: Historical Dimensions of Engagement and Disengagement* (1990); *Southern Africa: Civilizations in Turmoil* (1981); *Modern Africa, Change and Continuity* (1980), among numerous other works.

MINABERE IBELEMA is Associate Professor of Communication Studies at the University of Alabama at Birmingham. His research interests are in the press and democracy, press policies and professionalism, mass media and culture, and Africa's media image. He has published numerous journal articles and book chapters, including in *Africa's Media Image*, winner of the Society of Professional Journalists' Sigma Delta Chi Research Award in 1993. His non-academic essays have appeared in *Essence* magazine, the St. Louis *Post-Dispatch,* the Cleveland *Plain Dealer,* the *Christian Science Monitor*, and the Nigerian *Sunday Times*, among other publications.

OKECHUKWU C. IHEDURU is Assistant Professor of Political Science at James Madison College, Michigan State University. In addition to publishing several scholarly articles on Africa's political economy and maritime industry, he is the author of *The Political Economy of International Shipping in Developing Countries* (1996). Dr. Iheduru is a consultant to the World Bank on shipping policies of African states.

F. ABIOLA IRELE is Professor of African, French, and Comparative Literature at Ohio State University, with a joint appointment in the Department of Black Studies and the Division of Comparative Studies in the Humanities, and a courtesy appointment in the Department of French and Italian. Among other works, Dr. Irele is the editor of a volume, *Selected Poems of Léopold Sédar Senghor* (1977), and author of *The African Experience in Literature and Ideology* (1981; reprinted 1990). He is the general editor of the Cambridge Studies in African and Caribbean Literature.

ALI A. MAZRUI is the Director of the Institute of Global Cultural Studies and the Albert Schweitzer Professor in the Humanities at Binghamton University. He also holds the position of Albert Luthuli Professor-at-Large at the University of Jos in Nigeria, and is a Senior Scholar and Andrew D. White Professor-at-Large Emeritus at Cornell University. He has published innumerable books and articles on Africa and global culture, including *The World Federation of Cultures* (1980), a volume in the World Order Model Project.

EMMANUEL U. NNADOZIE is Professor of Economics at Truman State University (Missouri) and President of African Profiles International. His scholarly works have appeared in both academic and non-academic journals worldwide. He is the co-author of *Chad: A Nation in Search of Its Future* (1997).

FOLU FOLARIN OGUNDIMU is Associate Professor in the School of Journalism at Michigan State University. He has worked as a newspaper reporter, magazine editor, and senior editor for radio and television news in both the United States and Nigeria. His professional journalism experience includes assignment in Africa, Europe, and the United States. He is a former director of "Mass Media and Reporting in Britain." He is a past recipient of research fellowships from the Rockefeller Foundation in both developing and developed countries.

EBERE ONWUDIWE is Professor of Political Science and the Director of the National Resource Center for African Studies at Central State University (Ohio). His research focuses on aspects of anglophone Africa's political economy and Africa's image. He is the founding editor of *African Affairs Bulletin* and *The International Journal of African Studies*. He has published several scholarly articles and book chapters on Africa. His commentaries on African affairs have appeared in *Congressional Record–Senate*, the *Christian Science Monitor*, the *New York Times*, *International Herald Tribune,* and in many other U.S. and international newspapers and magazines.

TRACY D. SNIPE is Assistant Professor of Political Science at Wright State University with research interest in African and African American politics.